JONES AND BARTLETT SERIES IN BIOMEDICAL INFORMATICS
SERIES EDITOR JULES J. BERMAN

Clinical Information Systems

Perl Programming for Medicine and Biology
(© 2007, ISBN 978-0-7637-4333-8)
Jules J. Berman

Ruby Programming for Medicine and Biology
(© 2008, ISBN 978-0-7637-5090-9)
Jules J. Berman

Python for Bioinformatics
(© 2009, ISBN 978-0-7637-5186-9)
Jason Kinser

R for Medicine and Biology
(© 2010, ISBN 978-0-7637-5808-0)
Paul D. Lewis

Related Titles

Biomedical Informatics
(© 2007, ISBN 978-0-7637-4135-8)
Jules J. Berman

Exploring Bioinformatics: A Project-Based Approach
(© 2010, ISBN 978-0-7637-5829-5)
Caroline St. Clair and Jonathan Visick

*Medical Informatics 20/20: Quality and Electronic Health Records through Collaboration, Open Solutions,
and Innovation*
(© 2007, ISBN 978-0-7637-3925-6)
Douglas Goldstein, Peter J. Groen, Suniti Ponkshe, and Marc Wine

JONES AND BARTLETT SERIES IN BIOMEDICAL INFORMATICS
SERIES EDITOR JULES J. BERMAN

Clinical Information Systems
Overcoming Adverse Consequences

Dean F. Sittig

*University of Texas School of
Health Information Sciences at Houston*

Joan S. Ash

Oregon Health & Science University

JONES AND BARTLETT PUBLISHERS

Sudbury, Massachusetts

BOSTON   TORONTO   LONDON   SINGAPORE

World Headquarters
Jones and Bartlett Publishers
40 Tall Pine Drive
Sudbury, MA 01776
978-443-5000
info@jbpub.com
www.jbpub.com

Jones and Bartlett Publishers
Canada
6339 Ormindale Way
Mississauga, Ontario L5V 1J2
Canada

Jones and Bartlett Publishers
International
Barb House, Barb Mews
London W6 7PA
United Kingdom

Jones and Bartlett's books and products are available through most bookstores and online booksellers. To contact Jones and Bartlett Publishers directly, call 800-832-0034, fax 978-443-8000, or visit our website, www.jbpub.com.

Substantial discounts on bulk quantities of Jones and Bartlett's publications are available to corporations, professional associations, and other qualified organizations. For details and specific discount information, contact the special sales department at Jones and Bartlett via the above contact information or send an email to specialsales@jbpub.com.

Production Credits
Chief Executive Officer: Clayton Jones
Chief Operating Officer: Don W. Jones, Jr.
President, Higher Education and Professional Publishing: Robert W. Holland, Jr.
V.P., Sales: William J. Kane
V.P., Design and Production: Anne Spencer
V.P., Manufacturing and Inventory Control: Therese Connell
Publisher, Higher Education: Cathleen Sether
Acquisitions Editor: Molly Steinbach
Senior Editorial Assistant: Jessica S. Acox
Editorial Assistant: Caroline Perry
Production Assistant: Lisa Lamenzo
Senior Marketing Manager: Andrea DeFronzo
Composition: Toppan Best-set Premedia Limited
Cover Design: Scott Moden
Cover Image: © Sergei Didyk/ShutterStock, Inc.
Printing and Binding: Malloy, Inc.
Cover Printing: Malloy, Inc.

Library of Congress Cataloging-in-Publication Data
Sittig, Dean F.
    Clinical information systems : overcoming adverse consequences / Dean F. Sittig, Joan S. Ash.
        p. ; cm.
    Includes bibliographical references and index.
    ISBN-13: 978-0-7637-5764-9
    ISBN-10: 0-7637-5764-0
    1. Medical informatics.   2. Medicine—Information technology.   3. Medical errors—Prevention.
I. Ash, Joan.   II. Title.
    [DNLM: 1. Medical Order Entry Systems.   2. Attitude to Computers.   3. Equipment Failure.
4. Evaluation Studies as Topic.   5. Medical Errors—prevention & control.   WX 26.5 S623c 2010]
    R858.S555 2010
    610.285—dc22
                                                2009032655
6048

Printed in the United States of America
13 12 11 10 09   10 9 8 7 6 5 4 3 2 1

We dedicate this book to Cody Curtis, our project manager throughout the course of this work. Her steadfastness, organizational skills, encouragement, and ability to find humor in the most routine events provided us with the strength to see the project through to completion.

Brief Contents

About the Authors   xv
Preface   xvii
Acknowledgments   xviii

*Part I   Understanding the Unintended Adverse Consequences Associated with
          Clinical Information Systems   1*

Chapter 1   Computer-Based Provider Order Entry: Placing the Application
             into Perspective   3
Chapter 2   An Introduction to Unintended Consequences of Clinical
             Information Technology   29
Chapter 3   A Sociotechnical Assessment of Unintended Adverse
             Consequences Associated with Clinical Workflow   53
Chapter 4   Emotional Aspects of Clinical Information System Use   67
Chapter 5   Shifts in Power, Control, and Autonomy   77
Chapter 6   The Nature of Clinical Information System–Related Errors   85
Chapter 7   Clinical Decision-Support Systems   105
Chapter 8   Clinical Information Systems and Communication: Reciprocal
             Impacts   115
Chapter 9   Overdependence on Technology   125
Chapter 10   Persistent Paper: The Myth of "Going Paperless"   135

*Part II   Overcoming the Unintended Adverse Consequences Associated with
          Clinical Information Systems   145*

Chapter 11   Considerations for Successful CPOE Implementations   147
Chapter 12   The Importance of Special People   165
Chapter 13   Assessment of the Anticipated Consequences of CPOE Prior to
             Implementation   181
Chapter 14   Rapid Assessment of Clinical Information System Interventions   191
Chapter 15   Basic Microbiologic and Infection Control Information to
             Reduce the Potential Transmission of Pathogens to Patients via
             Computer Hardware   199
Chapter 16   Lessons from "Unexpected Increased Mortality After
             Implementation of a Commercially Sold Computerized Physician
             Order-Entry System"   215

Index   223

Contents

About the Authors   xv
Preface   xvii
Acknowledgments   xviii

Part I   Understanding the Unintended Adverse Consequences Associated with
         Clinical Information Systems   1

Chapter 1   Computer-Based Provider Order Entry: Placing the Application
            into Perspective   3
            1.1   Examples of CPOE Implementation Efforts   4
            1.2   Rationale for Direct CPOE   7
            1.3   Sociologic Barriers to Direct CPOE   11
            1.4   Logistic Challenges Involved in CPOE Implementation   14
            1.5   System Design Issues   17
            1.6   Use of Expert Systems to Facilitate CPOE   21
            1.7   Implications for Future CPOE Systems   23
            1.8   References   24

Chapter 2   An Introduction to Unintended Consequences of Clinical
            Information Technology   29
            2.1   Key Points   29
            2.2   Introduction   29
            2.3   Background   30
            2.4   Major Categories of Unintended Consequences Identified   30
            2.5   Discussion   45
            2.6   Conclusion   48
            2.7   References   48

Chapter 3   A Sociotechnical Assessment of Unintended Adverse
            Consequences Associated with Clinical Workflow   53
            3.1   Key Points   53
            3.2   Introduction   53
            3.3   Key Findings   55
            3.4   Conclusion   62
            3.5   References   65

ix

Chapter 4   Emotional Aspects of Clinical Information System Use   67
    4.1   Key Points   67
    4.2   Introduction   67
    4.3   Ways in Which Emotions Lead to Unintended
            Consequences   68
    4.4   Discussion   72
    4.5   Conclusion   74
    4.6   References   75

Chapter 5   Shifts in Power, Control, and Autonomy   77
    5.1   Key Points   77
    5.2   Introduction   77
    5.3   How Does CPOE Change the Power Structure in
            Organizations?   78
    5.4   Discussion   82
    5.5   Conclusion   83
    5.6   References   84

Chapter 6   The Nature of Clinical Information System–Related Errors   85
    6.1   Key Points   85
    6.2   Introduction   85
    6.3   Background and Methods   87
    6.4   Kinds of Silent Errors   88
    6.5   Discussion and Conclusion   96
    6.6   References   99

Chapter 7   Clinical Decision-Support Systems   105
    7.1   Key Points   105
    7.2   Introduction   105
    7.3   Results   105
    7.4   Discussion   110
    7.5   Conclusion   111
    7.6   References   113

Chapter 8   Clinical Information Systems and Communication: Reciprocal Impacts   115
    8.1   Key Points   115
    8.2   Introduction   115
    8.3   Key Findings: Reciprocal Impacts of CPOE on
            Communication   115
    8.4   Disturbance of Doctor–Nurse Communication: "There's Not
            That Physical Presence"   116
    8.5   Discussion   121
    8.6   Conclusion   122
    8.7   References   123

Chapter 9   Overdependence on Technology   125
    9.1   Key Points   125
    9.2   Introduction   125
    9.3   Results   126
    9.4   Discussion and Suggestions for the Future   129
    9.5   Conclusion   131
    9.6   References   133

Chapter 10   Persistent Paper: The Myth of "Going Paperless"   135
    10.1   Key Points   135
    10.2   Introduction   135
    10.3   Rationale for Persistent Paper   136
    10.4   Discussion   139
    10.5   Conclusion   141
    10.6   References   142

Part II   Overcoming the Unintended Adverse Consequences Associated with Clinical Information Systems   145

Chapter 11   Considerations for Successful CPOE Implementations   147
    11.1   Introduction   147
    11.2   Consideration 1: Motivation for Implementing CPOE   147
    11.3   Consideration 2: Foundations Needed Prior to Implementing CPOE   148
    11.4   Consideration 3: Costs   150
    11.5   Consideration 4: Integration/Workflow/Healthcare Processes   151
    11.6   Consideration 5: Value to Users/Decision-Support Systems   152
    11.7   Consideration 6: Vision/Leadership/People   154
    11.8   Consideration 7: Technical Considerations   156
    11.9   Consideration 8: Management of Project or Program/Strategies/Processes from Concept to Implementation   158
    11.10   Consideration 9: Training/Support/Help at the Elbow   160
    11.11   Consideration 10: Learning/Evaluation/Improvement   161
    11.12   Final Thoughts   162

Chapter 12   The Importance of Special People   165
    12.1   Key Points   165
    12.2   Introduction   165
    12.3   Special People   166
    12.4   Administrative Leadership Level   167
    12.5   Clinical Leadership Level   170

12.6   Support Staff Level: Bridgers   173
12.7   Discussion   176
12.8   Conclusion and Recommendations: New Bottles, New Wine   178
12.9   References   179

Chapter 13   Assessment of the Anticipated Consequences of CPOE Prior to Implementation   181
13.1   Key Points   181
13.2   Introduction   181
13.3   Background   182
13.4   Methods   183
13.5   Results   183
13.6   Discussion   185
13.7   Conclusion   189
13.8   References   189

Chapter 14   Rapid Assessment of Clinical Information System Interventions   191
14.1   Key Points   191
14.2   Introduction   191
14.3   Selection of Methodological Approaches   192
14.4   Development of the Field Manual   193
14.5   Preparation for Site Visits   194
14.6   Subject Selection   194
14.7   Data Collection   195
14.8   Data Management   195
14.9   Data Analysis   195
14.10   Results   196
14.11   Lessons Learned about Methods   196
14.12   References   196

Chapter 15   Basic Microbiologic and Infection Control Information to Reduce the Potential Transmission of Pathogens to Patients via Computer Hardware   199
15.1   Introduction   199
15.2   Review of Pertinent Microbiologic Concepts and Findings   199
15.3   Potential Solutions with Emphasis on Basic Infection Control Principles   205
15.4   Practical Applications of Infection Control Principles in Medical Settings   207
15.5   Summary   209
15.6   References   210

Chapter 16   Lessons from "Unexpected Increased Mortality After Implementation of a Commercially Sold Computerized Physician Order-Entry System" 215

    16.1   Introduction 215

    16.2   What We Can Learn from This Study 216

    16.3   Conclusion 219

    16.4   References 220

Index 223

About the Authors

Dean F. Sittig, MS, PhD

Dean F. Sittig is an associate professor at the School of Health Information Sciences at the University of Texas Health Science Center at Houston and a member of the University of Texas–Memorial Hermann Center for Healthcare Quality and Safety. Dr. Sittig earned a masters degree in biomedical engineering from Pennsylvania State University in 1984 and a PhD in medical informatics from the University of Utah in 1988. In 1992, he was elected as a fellow in the American College of Medical Informatics.

Dr. Sittig's research interests center on the design, development, implementation, and evaluation of all aspects of clinical information systems. In addition to Dr. Sittig's work on measuring the impact of clinical information systems on a large scale, he is working to improve our understanding of the factors that lead to success, as well as the unintended consequences associated with computer-based clinical decision support and provider order entry systems.

He has just finished co-authoring an award-winning book on clinical decision support, titled *Improving Outcomes with Clinical Decision Support: An Implementer's Guide*.

Finally, he is the founding editor of both The Informatics Review (www. informatics-review.com), an online serial devoted to helping clinicians and information system professionals keep up to date with the field of clinical informatics, and The ClinfoWiki (www.clinfowiki.org), an interactive, collaborative, online clinical informatics reference resource.

Joan S. Ash, PhD, MLS, MS, MBA

Joan S. Ash is professor and vice-chair of the Department of Medical Informatics and Clinical Epidemiology in the School of Medicine at Oregon Health and Science University (OHSU), Portland, Oregon. She holds masters degrees in library science, health science, and business administration, and a PhD in systems science. Dr. Ash has served on the boards of directors of the American Medical Informatics Association and the Medical Library Association, and on NLM's Biomedical Library and Informatics Review Committee. She is presently chair of the Board of Scientific Counselors for the Lister Hill National Center for Biomedical Communications of NLM. She is also an elected fellow of the American College of Medical Informatics.

Dr. Ash's research focuses on behavioral and social issues related to implementing clinical information systems, specifically computerized provider order entry (CPOE) and clinical decision support (CDS), and the use of qualitative methods for conducting such studies. She leads a team of researchers, the Provider Order Entry Team (POET), which has conducted national surveys of CPOE use and its unintended consequences and fieldwork in fifteen organizations to investigate success factors and unintended consequences of CPOE and CDS. Results are available on the POET web site at www.cpoe.org.

Preface

The publication of the Institute of Medicine's treatises on the state of healthcare quality and safety in the United States in the late 1990s and early 2000s has put incredible pressure on healthcare organizations of all sizes and types, but especially hospitals, to begin implementing state-of-the-art clinical information systems. After former President George W. Bush's pronouncement in January, 2004, that all Americans should have electronic health records within the next ten years, the pressure only increased. More recently, several scientific publications described unintended adverse events that occurred following implementation of these complex, state-of-the-art clinical information systems.

The goal of this book is to help those charged with the challenging task of implementing one of these clinical information systems within their own organization to better understand and begin to deal with the inevitable, unintended, adverse events that may occur. It is based on a multi-year academic research project carried out by the Provider Order Entry Team (POET) from the Oregon Health and Science University in Portland, Oregon, funded by the National Institutes of Health National Library of Medicine, and led by Joan S. Ash. In brief, this research project involved a dedicated, multidisciplinary team of investigators, including the authors of this volume, who traveled around the United States visiting a number of carefully selected pioneering healthcare systems with advanced clinical information systems. At each site, the group conducted extensive interviews with key clinical, technical, and administrative informants as well as carrying out hours upon hours of observations of clinicians as they used, and often struggled with, these clinical information systems.

The history of this unintended consequences project has its roots in a prior study of success factors for the implementation of computerized provider order entry (CPOE). During that project, the authors naturally encountered barriers to success while probing for positive strategies. Even the organizations that have the best systems have stumbled along the way; fortunately representatives from within those organizations have shared stories about the problems so that lessons can be learned from them. Although excellent organizations do their best, they cannot always avoid or overcome certain unintended consequences that come their way during any implementation. This preliminary knowledge about unintended consequences led to the study that is the focus here.

The methods used for the unintended consequences study are ethnographic, including primarily qualitative methods, and also small surveys. Sites were purposively selected with the help of experts to provide a variety of sizes and types. Informants were carefully selected based on their roles. The team, composed not only of the authors, but also of clinicians and informatics specialists, spent several days at each site, after gathering information about the history of the system and the system itself. The team conducted semi-structured interviews with information technology, clinical, and administrative staff that were knowledgeable about CPOE. Team members shadowed and conducted informal interviews with clinicians as they worked, writing detailed field notes that were later expanded to be analyzed by the team. A categorization of types of unintended adverse consequences was the result, and the data provided quotes and examples so that for each type the team could delve deeply into its nature. Although "clinical decision support systems" (CDS) did not emerge as a type, since it is often a cause of rather than a kind of unintended consequence, a chapter about CDS is included here because it offers guidance about overcoming CDS-related unintended consequences.

Each chapter in this monograph is based on a published or as-yet unpublished paper written by this team, though each has been heavily edited so that repetition about, for example, methodology, is truncated or omitted. Several of the papers stem from the earlier study; these chapters about communication, special people, and the emotional aspects of CPOE present results that were subsequently validated during the unintended consequences study.

The book begins with an overview of the various types of unintended adverse consequences that often occur following implementation of these systems, and the next eight chapters describe them in detail. Chapter 10, at the end of Part I, summarizes the issues that need to be addressed for an implementation to be successful, with as few unintended consequences as possible. At the end of each chapter, there is a list of questions that you can use to help assess how your organization is doing with respect to each of these problematic areas.

Part II of the book begins with a list of considerations that organizations should keep in mind during the implementation and optimization process. Chapter 12 describes the role of special people within your organization who can help ensure that clinical information systems work as planned. The next chapter describes a relatively simple assessment tool that can be used to gauge an organization's readiness to deal with these unintended consequences. The book ends with a description of an innovative qualitative research methodology that can be used to study one's own organization.

In sum, this book is designed to both warn and inform about what often happens following implementation of a state-of-the-art clinical information system. In addition, it provides the knowledge and tools to help an organization identify, measure, and then overcome these obstacles and utilize clinical information systems to achieve the high-quality, cost-effective, efficient, safe, timely, and patient-centered healthcare system that the Institute of Medicine has envisioned.

Acknowledgments

We would like to acknowledge the tremendous insight, hard work and dedication of the various members of the Provider Order Entry Team (POET) from the Oregon Health and Science University in Portland, Oregon. Over the years, this team has included:

Emily Campbell, Cody Curtis, Jim Carpenter, Richard Dykstra, Lara Fournier, Paul Gorman, Bill Hersh, Melissa Honour, Ken Guappone, Mary Lavelle, Jason Lyman, Carmit McMullen, Josh Richardson, Veena Seshadri, Zoe Stavri, and Adam Wright. Without this multidisciplinary team, we could never have achieved so much. We are indebted to the National Library of Medicine for providing financial support to this team since 2000. We would also like to thank Jules Berman, series editor of the Jones and Bartlett Series in Biomedical Informatics, and the editorial staff members at Jones and Bartlett who gave us constant encouragement, exceptional copyediting, and a "push" when we needed it. In addition, the American Medical Informatics Association deserves thanks for allowing us to include a number of journal and proceedings papers from its publications in this volume. Finally, we would like to thank our spouses, JoAnn and Paul, who have supported us over the years and enabled us to travel all over the country to conduct our studies.

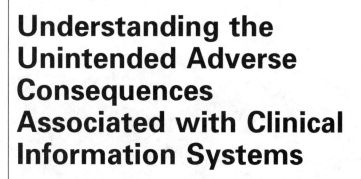

Understanding the Unintended Adverse Consequences Associated with Clinical Information Systems

1 Computer-Based Provider Order Entry: Placing the Application into Perspective

Based on: Sittig DF, Stead WW. Computer-based physician order entry: the state of the art. *J Am Med Inform Assoc.* 1994;1(2):108–123.

Time is precious. Neither a sick patient nor a busy clinician has a moment to spare. Errors can be disastrous. An incorrect medication or a high-risk, invasive test performed on the wrong patient could prove life-threatening. Money is limited. Any method of eliminating unnecessary steps must be exploited. Communication—transferring the appropriate information to the correct person at the time and place it is required—is vital.

Perhaps the major impact of information management technology on modern society is the way it is changing the manner and ease with which we communicate. For more than 20 years, computer-based, direct physician order entry (CPOE) has been put forth as a potential way to improve communication within the healthcare process. As early as 1970, when Collen listed several of the general objectives of a medical information management system, he included "to communicate patient data from professionals providing medical care (doctors, nurses, technicians, etc.) into the patient's computer-based medical record, to other professionals (e.g., dietitians), and to hospital services (e.g., radiology)."[1] He specifically stated, "Physicians should enter medical orders directly into the computer" as a means of ensuring quality.

While the logic of eliminating the middleman through CPOE is easy to comprehend, actual implementation of CPOE is more difficult than one might imagine.[2–7] A small number of institutions have had success, but the vast majority of institutions that attempted to implement CPOE, as well as corporations that attempted to sell CPOE systems, during the 1970s and 1980s met with failure of varying degrees. For example, by 1982, Spectra, in conjunction with its hospital clients, had spent more than $200 million to create what was only a rudimentary order-entry system when development was halted.[8] Thus, during this 20-year period, CPOE was pursued only at truly pioneering institutions.

The late 1980s and the early 1990s saw a renewed emphasis on CPOE owing to several important factors.[9,10] First and foremost among these developments were advances in information management technology, coupled with the concomitant decreases in price that permitted new approaches to the user interface.[11,12] Second, a significant increase in general computer literacy occurred within the medical profession and among hospital administrators.[13] Finally, the interest in developing a complete computer-based patient record to increase quality of care and documentation of services, while reducing costs, has brought the need for direct interaction between practitioners and databases to the attention of medical personnel.[14,15]

Experience has shown that it is difficult to implement CPOE successfully. One may ask, "Should CPOE be a goal?" and "Can CPOE now be achieved outside pioneering institutions?" In this chapter, we attempt to answer these questions by presenting examples of CPOE implementation efforts. Next, the rationale for, and barriers to, CPOE are examined. We then identify key system design issues. The sources for this chapter include the peer-reviewed scientific literature and published technical reports. Thus negative results may be underrepresented. Only those aspects relevant to direct entry of orders by the physician, as distinguished from systems to process orders entered by a third party, are included.

1.1 Examples of CPOE Implementation Efforts

An important distinction must be made regarding the fundamental approaches taken by different developers of CPOE systems. The first attempts, such as those by Technicon Data Systems (previously called TDS Healthcare Systems Corporation, Alltel, and more recently Eclipsys), were intended to develop hospital information systems for use by clinicians. They identified CPOE as one of the primary objectives at the outset. A second group of investigators focused early efforts on developing a computer-based patient record or on clinical decision support systems. Direct order entry was then added as an extension. The first group of developers focused on implementing straightforward CPOE for inpatients house-wide, whereas the second group explored more elaborate functionality for small classes of users, in a mixture of outpatient and inpatient settings. Examples of these development approaches are described next.

1.1.1 Technicon Data System (TDS)

In June 1971, the National Center for Health Services Research selected El Camino Hospital, located in Mountain View, California, to demonstrate and evaluate the Technicon Medical Information Management System.[16] TDS is a hospital information system designed for use by nurses, physicians, and other healthcare workers with a goal of expediting the healthcare process. It was developed and tested on a single nursing station during the 5 years prior to 1971. At the time, El Camino was a 464-bed, general community hospital serving patients under the care of their private physicians. It did not have an internship or residency program. It averaged more than 22,500 admissions per year between 1970 and 1972, with an average length of stay of 5.4 days.

TDS was implemented in phases. The admission department began to use the system in December 1971.[16] In January 1972, the first nursing unit came online for order entry and reporting of results. All nursing stations were activated by October 1972. By October 1974, 78% of the physicians used the system for either entering orders or reviewing results, and 45% of all orders were entered directly by physicians.[16]

Objective studies of the accuracy and completeness of the computer-based order-entry system at El Camino proved CPOE to be beneficial. Following implementation of CPOE, errors of omission in medication orders concerning site and route of administration (7.9% initially) and dosage scheduling (1.3% initially) dropped to less than 0.5% ($p < 0.01$). Inclusion of clinical indications for radiologic examinations and electrocardiographic monitoring increased significantly from less than 4% to more than 35% of orders examined.[16]

A subjective assessment of the system showed that the physicians who used the system the most were its strongest proponents. This raised the still unanswered question: Do they use the system because they like it, or do they like the system because they use it? In general, those physicians who had adopted personal order sets were the most enthusiastic supporters of the system.[17]

These findings were confirmed in another early TDS installation that took place at the New York University Medical Center. In a pre/post-installation study conducted there, investigators demonstrated a 22% reduction in physician medication orders that omitted the site, route of administration, dose, or schedule, and a 32% reduction in radiology orders that omitted the clinical indication. In addition, this facility reduced its departmental order turnaround times by 4.9 hours (71%) in the pharmacy and by 2.4 hours (9%) in the chemistry laboratory.[18]

More recently, the experience at the University of Virginia (UVA) has confirmed the difficult course that CPOE follows, even with ultimate success. The UVA began installation of a CPOE system in 1985 with basic administrative functions such as admission, discharge, and transfer. Beginning in 1988, radiology and dietary came online. Mandatory CPOE for radiology, dietary, and laboratory use followed, accompanied by result reporting. Pharmacy order entry followed in July 1989 but met with strong opposition from the house staff. Dissatisfaction with the system peaked a year later, when a work action was initiated by a group of the most frustrated residents. During an open meeting with these physicians, senior members of the medical center administration stressed the system's strategic importance and reaffirmed their decision to keep it operational. Following this meeting, the work action was stopped and a few days later the new house staff class arrived and was oriented to the system with little difficulty.[19] Ordering from the remaining ancillary services and nursing procedures was introduced in the final phase. The entire project took 3 years longer than planned and cost nearly three times the original estimate. The end result, however, has been good; house staff now tell recruits that the CPOE system is one of the reasons to come to UVA.

Massaro stated that the UVA medical information system (MIS) did not truly begin to be integrated into the operational culture of the institution until after the Computer and Information Sciences Executive Committee was created and began to meet weekly to address these important changes.[7] This committee was composed of the chairs of medicine, surgery, and pediatrics; the executive director of the medical center; the director of nursing; the chief information officer; and the senior associate vice-president of the UVA Health Sciences Center. Perhaps the most important lesson learned at UVA is that information technology alone cannot fix many problems the technology did not create, but the technology can accentuate existing problems by forcing strict adherence to seldom-followed rules and by diverting attention from the fundamental issues involved.

1.1.2 The HELP Clinical Information Management System

At the LDS Hospital in Salt Lake City, Utah, Gardner et al.[20] developed an online critiquing system for use by physicians and nurses when ordering blood products. Their system utilizes the HELP clinical information management system[21-23] to integrate patient data from eight sources within the hospital:

- Blood bank (e.g., blood that is already ordered)
- Admit/discharge/transfer (e.g., patient demographics)
- Clinical laboratory (e.g., hematocrit and hemoglobin values from the complete blood count [CBC])
- Surgical schedule
- Nurse charting (e.g., vital signs—heart rate, blood pressure, and fluid loss)
- Bedside monitors in the intensive care unit (ICU) (e.g., blood pressure, O_2 saturation)
- Blood-gas laboratory (e.g., hemoglobin, O_2 saturation)
- Physicians' and nurses' data entry (e.g., current bleeding status plus reasons for overriding the system)

For the academic year 1988–1989, there were 13,082 orders for 48,581 units of blood products (only 29,702 were actually issued) entered into the HELP system.[24] Physicians entered 42.7% of the orders directly from the terminals. Standing orders, which are automatically initiated from the computerized surgical schedule for 21 specific procedures (e.g., open-heart surgery or total hip replacement) accounted for another 8.1% of the total orders. The remainder of the orders were entered by nurses and consisted of 27.4% written orders (38 of the 405 physicians—9.4%—accounted for 66.6% of the written orders), 14.1% verbal orders, and 7.8% phone-in orders. The average time to place an order using the computer system was 2.2 minutes. Because the information from the clinical laboratory and previous blood orders were presented during the order-entry process, Gardner et al. concluded that the computer-based system was time- and cost-effective compared with the time spent searching the manual, paper-based chart for the latest laboratory results and previous blood orders.

Given that the main goal in introducing the direct CPOE system for blood products was to improve compliance with Joint Commission on Accreditation of Healthcare Organizations (JCAHO) guidelines, Lepage et al. performed a quality assurance review of all orders placed during the 1988–1989 academic year.[24] Of the orders entered, 86.8% (81.3% of the actual units ordered) complied with criteria established by the medical staff (criteria were derived from the medical literature with slight modifications due to the altitude, 4600 feet above sea level, of the LDS hospital). Of the 13.1% of orders that initially appeared not to meet the established criteria, 27.5% occurred because "no reason" was specified in the order (3.6% of the total). Nurses were specifically instructed not to "guess" at reasons for order exceptions (e.g., those not meeting the established criteria based on data already in the database) on written or verbal orders without specific reasons stated. After careful review, of the remainder of the orders not meeting established criteria, the quality assurance department found that only 48 (0.37% of the total number of orders) were true exceptions to the established criteria for utilization of blood products.

1.1.3 Regenstrief Medical Center

Investigators at the Wishard Memorial Hospital began using the Regenstrief Medical Record System in 1973.[25] Since 1984, they have developed successive components of a CPOE system.[26] An innovative aspect of their work was the inclusion of CARE rules (patient-specific medical reminders based on algebraic combinations of raw data and/or other CARE rules), which are executed automatically when triggered by data or order entry for the associated test or treatment.[27]

In addition, McDonald and colleagues have developed a "medical gopher," which assists in the performance of many routine clinical activities by reducing the work of locating pertinent general medical information. For example, when a physician orders treatment for a specific problem, the medical gopher displays the most common workup and/or treatment for the problem. Displayed items can then be ordered using one or more menu selections. Drug–drug, drug–test, and drug–diagnosis interactions are also detected automatically, and a warning message is displayed before the order is completed.

In the early implementation phase of the CPOE system, all data entry was accomplished via keyboard. Using this early technology (computers equipped with 80,286 processors and keyboards) McDonald and Tierney found average order times of 30 seconds/test.[26] They subsequently upgraded their PCs to 80,386 processors and added mouse-based data entry but still reported that physicians prefer the keyboard data entry over the mouse in a ratio of ten to one.[28] Because the system includes data display for decision support in the order process, computer-based CPOE requires more time than the manual system (58.5 versus 25.5 minutes during a 10-hour period, $p < 0.0001$).[29] Of greater importance was the fact that they were able to reduce inpatient charges and estimated hospital costs significantly.[29] In a subjective evaluation conducted by Tierney et al., 70% of the responding clinicians indicated that using CPOE made their work more interesting, and 44% believed that their work was done faster using the system. Interestingly, 52% said that the CPOE made their work easier.[29]

1.2 Rationale for Direct CPOE

Physician order entry is a strategic option for facilities considering how to deal with the twin requirements of health reform: improvement and documentation of quality while containing costs.[30] Although it is easy to understand the importance of CPOE from the perspective of organizational leadership, it is more difficult to rationalize use of such systems from the perspective of the individual worker. For CPOE to be successful, project leaders must carefully consider the work patterns of individuals. CPOE systems must deliver tangible benefits to users if the change is to be made without creating a net disadvantage for any key group.

Benefits that can accrue from implementation of CPOE can be categorized as follows: process improvement, cost-conscious decision making, clinical decision support, and optimal use of physician time. Process improvement involves the reengineering of the entire order-entry process so that those responsible for the decisions are directly involved in the entering of orders. This improves the conveyance of orders as well as enabling the system to provide real-time feedback to clinicians regarding the appropriateness of certain orders.[30] Cost-conscious decision making involves presenting clinicians with less expensive tests and/or treatments for a specific diagnosis, or presenting the cost of each test or treatment to the clinician at the time it is ordered.[29,31] Clinical decision support involves providing information relevant to formulation of a diagnostic hypothesis or appropriate therapy. Benefits of one type often cause gains in the other areas. For example, process improvement also decreases fragmentation of the health professional's time.

1.2.1 Process Improvement

The process improvements identified in this section save the ordering physician time by decreasing the need to repeat tests or procedures whose results are lost, or actions that have not been carried to completion. They improve the timeliness and reduce the cost of care.

Process improvement via CPOE can achieve the following goals:

- Eliminate lost orders because the initial record of an order is made directly into a computer database. Thus follow-up on overdue orders can be automated.

- Virtually eliminate ambiguities caused by illegibility of handwritten orders (although orders typed in as free text can still be potentially misunderstood). Incomplete orders are not possible. At the Kochi Medical School Hospital in Japan, inquiries from pharmacists to the ordering physicians after a prescription audit of handwritten prescriptions amounted to 11% of orders prior to the adoption of CPOE.[32] Following implementation of the hospital's computer-based order system, this percentage dropped to 1.4% (approximately one every 4 hours).

- Generate related orders automatically. For example, a heparin-flush order could be generated with every nursing order to establish an indwelling intermittent injection site.

- Generate automatic stop orders for prophylactic antibiotics.[33]

- Continuously monitor all orders for a particular patient and prompt the clinical staff regarding whether an order is a duplicate.[34]

- Reduce the time required to fill drug orders. For example, in a randomized controlled clinical trial, Tierney et al. were able to show that admitting drug orders were filled 63 minutes faster on average, while daily drug orders were filled 34 minutes sooner.[29]

- Integrate quality-assurance monitors into the order-capture process.[20] For example, at the LDS Hospital in Salt Lake City, all physicians are required to provide reasons for ordering blood products if the database (e.g., laboratory results, operative schedule) does not contain a justification based upon established criteria.

- Reduce the amount of money hospitals spend on preprinted multipart forms used in the order entry process. Hodge estimates that savings of $5000 per month or more are not unusual.[35]

1.2.2 Support of Cost-Conscious Decisions

Support of cost-conscious decision making involves increasing the awareness of the cost of a service so that it is avoided entirely when not needed or by identification of less expensive alternatives. Cost-conscious decision making through CPOE can achieve the following goals:

- Assist in keeping prescribing practices consistent with a hospital's established formulary, resulting in significant cost savings.[36]
- Educate physicians regarding cost-effective medication options. For example, Reynolds et al. found that presenting educational comparative therapeutic and cost data for clindamycin, cefoxitin, and metronidazole resulted in a 400% increase in the use of metronidazole with a concomitant decrease in clindamycin usage ($p < 0.001$).[37] Kawahara et al. showed that the percentage of patients with community-acquired pneumonia due to *Haemophilus influenzae* or gram-negative enteric rods who were prescribed cefuroxime decreased from 100% to 22% over a 1-year period (the average cost of antibiotics per patient decreased from $123 to $48)[38]; see Figure 1-1.

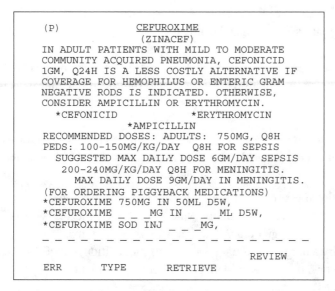

```
(P)                CEFUROXIME
                   (ZINACEF)
IN ADULT PATIENTS WITH MILD TO MODERATE
COMMUNITY ACQUIRED PNEUMONIA, CEFONICID
1GM, Q24H IS A LESS COSTLY ALTERNATIVE IF
COVERAGE FOR HEMOPHILUS OR ENTERIC GRAM
NEGATIVE RODS IS INDICATED. OTHERWISE,
CONSIDER AMPICILLIN OR ERYTHROMYCIN.
   *CEFONICID            *ERYTHROMYCIN
          *AMPICILLIN
RECOMMENDED DOSES: ADULTS: 750MG, Q8H
PEDS: 100-150MG/KG/DAY  Q8H FOR SEPSIS
   SUGGESTED MAX DAILY DOSE 6GM/DAY SEPSIS
   200-240MG/KG/DAY Q8H FOR MENINGITIS.
     MAX DAILY DOSE 9GM/DAY IN MENINGITIS.
(FOR ORDERING PIGGYBACK MEDICATIONS)
*CEFUROXIME 750MG IN 50ML D5W,
*CEFUROXIME _ _ _MG IN _ _ _ML D5W,
*CEFUROXIME SOD INJ _ _ _MG,

_ _ _ _ _ _ _ _ _ _ _ _ _ _ _ _ _ _ _

                             REVIEW
ERR       TYPE        RETRIEVE
```

Figure 1-1 The cefuroxime order-entry screen. Notice that the physician can select the cefonicid alternative directly from this screen without returning to the alphabetical index of medications. Selection of the cefonicid alternative will take the physician directly to the cefonicid dosage selection screen. This saves the physician several steps in the order-entry process.

Adapted from: Kawahara NE, Jordan FM. Influencing prescribing behavior by adapting computerized order-entry pathways. *Am J Hosp Pharm.* 1989;46:1798–1801.

- Prompt the ordering physician regarding optimal routes of medication administration. For example, if a patient is taking pills orally, and a new medication that exists in equally effective oral form is ordered to be given IV, then the system could provide the information that the oral form is available and costs significantly less.[39]

- Inform physicians of test charges before completion of the order. Tierney et al. showed that both the number and the cost of tests ordered by resident physicians could be reduced significantly by this simple action.[31]

- Reduce the number of tests ordered by presenting previous test results to clinicians before they order new tests. Specifically, Tierney et al. were able to demonstrate that resident physicians who were automatically shown past test results ordered 8.5% fewer tests, resulting in 13% lower charges as compared with physicians not shown such results automatically.[40] The same study documented that displaying the previous test results added only 4.5 seconds (8%) to the entire order-entry process.

- Reduce laboratory costs by presenting physicians with predictions of test abnormalities, based on predictive equations (derived from retrospective patient data) calculated using current patient specific data in the system. One hospital was able to reduce the number of tests ordered as well as the cost of clinical laboratory testing by 8.8% ($p < 0.05$). Reductions of more than 10% for the most commonly ordered tests (electrolyte levels and complete blood cell counts) were attributed to this technique.[41]

1.2.3 Clinical Decision Support

Many investigators argue that the greatest long-term benefit of CPOE will come from the integration of clinical decision support into the order-entry process. Clinical decision support requires a current and accurate knowledge base. The difficulties in developing and maintaining such a resource are significant.[42] Clinical decision support can be applied at several different levels of complexity. For example:

- The system can also make available to the clinician relevant information such as online laboratory manuals, the *Physician's Desk Reference,* a textbook of medicine, or even the entire Medline database.[26]

- Informational text can be inserted into the order-entry pathway, providing accessible and consistent information when the therapeutic decision is being made. This can reinforce guidelines (e.g., maximum potassium concentration that can be infused by a peripheral intravenous route). In addition, CPOE can be an effective method for rapidly disseminating drug-recall information.[36]

- The system can automatically make complex calculations for drug dosage and/ or parenteral and enteral feedings, based on patient age, sex, weight, and other clinical information.[43,44] Likewise, the system can check whether the dose, duration, and frequency of administration are within hospital guidelines.[34] For ethical, clinical, and medicolegal reasons, it is probably best if these calculations are used in an "open-loop" environment in which a physician always has the final decision.

- The system can generate reminders at the time a physician enters the order. Reminders can assist with the following situations:
  1. Interactions among concurrent drugs that may have been ordered by multiple practitioners (drug–drug interactions)
  2. Potential interactions between laboratory tests and specific drugs (e.g., a laboratory test whose results will not be meaningful because a particular drug has been given)
  3. Drug orders that should be modified based on laboratory values[45]
  4. Potential allergies[32]
  5. Potentially toxic conditions requiring attention (e.g., high drug levels in laboratory results; or the order of an overdosage; or the order of interacting drugs that increase levels)[46]
- The system can automatically suggest certain therapeutic orders for review by a physician before they become active, based on clinical data available within the system.[47,48]

1.2.4 Optimization of Physician Time

The physician gains satisfaction and competitiveness through the process-improvement and decision-support aspects of CPOE. When correctly implemented, CPOE should pay back the physician directly through time savings. In addition, CPOE delivers the following benefits in terms of optimization of physician time:

- Physicians can place orders from any location in the hospital (and some hospitals allow physicians to log in from their homes or offices) and be confident that they will be carried out in a timely and appropriate manner.[9]
- The number of telephone calls inquiring about orders can be greatly reduced.[49,50]
- Order sets can reduce the need for physicians to memorize and regurgitate routine orders, thereby freeing them to concentrate on identifying the unique features of a patient's illness and then tailoring the care plan to reflect those differences. Indiscriminate use of order sets can result in unnecessary orders, however.

1.3 Sociologic Barriers to Direct CPOE

Physician order entry is consistent with the industrial adage that data capture and data use should occur directly at the point of service. Achieving CPOE is difficult because in the healthcare system, the most highly trained and compensated personnel (e.g., physicians and nurses) are at the point of service, unlike other industries that have lower skilled and less costly personnel in those positions.[7] This difference leads to a common perception that the purpose of CPOE is to save money for the hospital by shifting work from clerks to physicians.[51] In fact, the hospital must be willing to invest money to give the physicians a better method for writing orders.

The goals of CPOE are to capture a non-ambiguous order at the source, to permit integration of decision support into order generation, and to act on orders in a more timely fashion. The institution must communicate clearly the strategic importance of CPOE and work with physicians and other care providers to develop an approach that they see as helping them as individuals. If this communication is not put in place early, distrust and fear will build powerful barriers to CPOE implementation.[7]

Another related fear is that CPOE will "de-intellectualize" the practice of medicine by making physicians use protocols. In reality, CPOE can decrease the amount of time that physicians spend on routine aspects of care, allowing them to focus their attention on the unique aspects of a case. In addition, development of a clinically defensible order set, based on data gathered about variations in practice patterns or the results of clinical trials, can be an educational experience for attending physicians and students alike.[52]

1.3.1 Changing Practice Patterns

Hospital information system (HIS) developers commonly build order-entry pathways, or menus, for clerical use. Such system developers obtain input from the ancillary departments rather than from physicians.[53] With a paper order sheet, when a physician writes an order in conflict with the hospital's or a department's rules, a clerk may buffer the physician from the problem by modifying the order during the data-entry process. Otherwise, the department will later call the physician to resolve the problem. The clerk learns individual physician preferences over time, often through repeated criticism when things go wrong. If the physician changes patient care units, or a key staff member leaves, then this protective buffer provided by the clerk must be redeveloped.

With CPOE, there is no buffer between the physician and the institution's rules/procedures. The developers and physician users must put energy into training the computer system so that a "human buffer" is not necessary. This investment must be repeated if the CPOE system undergoes major redesign.

Physician order entry forces physicians to modify their established practice routines and work-flow patterns. CPOE is best introduced by including physicians in the entire decision-making process. In hospitals with house officers, it is difficult to involve the interns (1-year postgraduate MDs), who will be the main users of a CPOE system, in the developmental decision-making process. Interns, of course, have very little free time. While residents (2- to 4-year postgraduate MDs) may have more time, the CPOE selection and implementation process may be longer than the remainder of their residency program. The residents who have been involved in the decisions will not be around when the system is finally put into use. Subsequent generations of residents may perceive that they were not consulted in the selection process and, therefore, want nothing to do with the system.[54] Conversely, residents who come to the hospital after CPOE has been implemented may perceive the system as the status quo and raise very few objections. In several instances, a CPOE system has become a "strategic advantage" in recruiting new residents.

In an attempt to minimize the disruptions to the traditional order-entry procedure during the implementation of CPOE, the Albany Medical Center offered a transcription service (to enter orders into the system for the physician) from 10:00 A.M. to 6:00

P.M. In a review of 1 month's activities, 73% of the orders (4836 total orders] were entered by the physicians themselves; an additional 9% were verbal orders entered by the nursing staff. Developers attributed the limited usage of the transcription service (18% of all orders entered) to problems with the transcriptionist. Specifically, they mentioned problems with reading physicians' handwriting, delays in entering the orders, and the fact that someone was expected to verify the data entered. Many of these problems were the same problems that CPOE was designed to address.[55]

Massaro, in an attempt to alleviate the perceived CPOE-related increase in clerical work required of the house staff, used a fax machine to send handwritten orders to the pharmacy on three separate patient care units.[19] Initially 22% of the orders were captured in this manner. After 3 months of operation, as problems with the CPOE system were resolved and the users gained experience entering orders, use of the fax system dropped to 2% to 3% of the total order volume. Subsequently, the fax machine was eliminated. Others have commented that use of the fax machine to send physicians' orders to the pharmacy does not address many vital issues.[56,57]

1.3.2 Shifting Roles within the Care Team

Physician order entry allows order entry to take place from anywhere within the institution and possibly from home. Accordingly, CPOE should decrease the need for verbal orders. Indeed, care must be taken to keep verbal orders from becoming a way of avoiding use of the CPOE system. At the same time, policy should permit verbal orders when they make sense. For example, if a physician is called during the night and told that "a patient cannot keep down his Compazine," the physician should be able to ask the caller to change the administration route to rectal from oral (without having to log into a system). Nurses might view entry of verbal orders as clerical work for the physician and refuse to accommodate them. Achieving the correct balance will require thoughtful discussion between members of the care team.

1.3.3 Changes in Traditional Teaching Patterns

Physician order entry will change the way people learn. In the past, clinical clerks (students) have learned by repetitively writing orders. Order sets, which can be part and parcel of efficient CPOE, eliminate this approach. At the same time, ready access to information such as decision trees or clinical simulations during the order-entry process should provide new and better ways of teaching students and trainees.

Physician order entry can also alter the manner in which students write orders. Orders from students ordinarily need to be discussed and cosigned by a resident. At the University of Virginia, residents found that it consumed more time to correct computer-based orders input by medical students than to input correct orders themselves. Therefore, the medical students received less of the resident's time and teaching.[7] Nevertheless, Tierney et al. found that medical students value the "educational extras" available on the Regenstrief CPOE system. Students felt more at ease writing orders using the Regenstrief CPOE than using paper because it allowed them to write orders without "defacing" the chart and/or making their mistakes permanent.[58]

A concern has been raised about what will happen when the physician who has been trained using CPOE leaves and goes to another institution.[6] First, all learning is site-specific to some extent. This problem has not been dealt with in any systematic manner to date. Second, if information technology is used to teach physicians how to make decisions about what to do, they may be able to pick up new routines more rapidly. Third, information technology is becoming ubiquitous. Digital information resources such as Medline,[59] COACH,[60] DxPlain,[61] and QMR[62] are becoming available to practitioners, however remote, via personal computers or telephone.

1.3.4 Inadequate Institutional Policies

Implementation of CPOE can force strict and literal interpretation of the policies, rules, and procedures of a medical center. This can cause frustration to users.[55] Policy problems may not be created by a new computer system, but they will most often be brought to light by its implementation. Policies and rules in place may not be followed rigorously.

Implementation of CPOE cannot proceed until procedural conflicts are resolved. Therefore, resolution of these problems must be handled rapidly and effectively at the highest levels within the organization (e.g., in an academic medical center by the dean, the director of the hospital, and the chief information officer, or in a private group practice by the partners)—an approach most organizations are not used to following.

1.4 Logistic Challenges Involved in CPOE Implementation

The logistic challenges surrounding CPOE implementation often significantly outweigh the technologic challenges. Logistic problems include the plan for phasing in each step of implementation, training issues, the number of terminals required, and the time required to enter and review data. The hospital must commit sufficient capital, system personnel, and support personnel to achieve the transition. The physicians must commit their time and adapt their practices somewhat if the organization is to achieve better order generation and management.

1.4.1 Implementation Phasing

Phasing the stages of CPOE implementation is one of the most difficult aspects of installing a new CPOE system.[54] Possibilities include the following options:

- Implement all functions at all locations at one time throughout the hospital.
- Implement all functions at one location. When that location has stabilized, begin rolling out the system to the remaining locations, one at a time.
- Implement limited functionality at all locations (e.g., laboratory test ordering). As these functions are accepted into the normal clinical routine, begin to roll out more functions.

- Combinations of these approaches, such as implementing limited functionality at one location, are also possible. When that functionality has stabilized, one can begin rolling out more functions at that location, and then roll out the system to other locations throughout the hospital.

There are many problems with implementing a system in stages, or in one portion of the medical center and not others. Problems include (1) handling patient transfers from a "computerized" unit to a paper-based unit; (2) movement of clinicians between units on which the system is used and ones on which it is not; (3) dealing with two types of order processing in ancillary departments; and (4) converting the paper record to the computer-based system, and vice versa, as the patient is moved. The transition period can be traumatic to personnel and can generate unpredictable crises and costs.[15]

McDonald et al.[28] state that introducing CPOE in an outpatient clinic is easier than attempting the same feat on an inpatient ward. These researchers found that for inpatients, the number of orders written and their variety were two orders of magnitude greater than in the outpatient clinic. In addition, telephone orders, student orders (which must be cosigned), and negotiable transfer orders that take effect if the bed is needed by a sicker patient added complicating wrinkles to the inpatient order-entry problem. Moreover, the turnover rate at McDonald's institution was higher (nearly 50 new physician/medical student users every 6 weeks) in the inpatient setting, which resulted in higher training costs because the entire hospital was not using the system.[28]

Schroeder[36] described a successful transition period in which no more than one patient care unit was brought online at a time. During the year-long implementation phase, the facility brought an average of one new unit online every 2 weeks. Specially trained pharmacy technicians were available 24 hours per day for the first 3 to 4 days, and waited unobtrusively until signs of frustration became apparent; they were instructed to offer their assistance at that point. In this way, the number of truly negative experiences was greatly reduced.

1.4.2 Number of Terminals and Location

Ideally, a workstation would be available wherever a physician might enter an order. In reality, the high-powered workstations required to support CPOE are too expensive to place wherever they might be needed. Therefore, CPOE currently requires use of a pool of shared workstations distributed around the patient care unit. Many institutions using CPOE have between three and five terminals per nursing station or approximately one terminal for every five to ten patients (depending, in part, on patient acuity).[7,29,36]

The number and location of workstations must be selected so as to minimize the number of times that physicians have to wait for a terminal and the distance that they have to walk. Even so, physicians may help by staggering rounding patterns where possible and by using clipboards to note orders prior to subsequent entry. In the future, hand-held tablets that communicate back to host databases over wireless networks may allow physicians to enter orders from anywhere within the hospital. For now, the pressure for better user interfaces has resulted in order-entry systems that require more computer power than can be managed today with a truly hand-held device.

1.4.3 Training Users

Installing a CPOE system requires massive training efforts. One must factor in not only the cost of the training staff, but also the cost of adding more workers to maintain hospital functions while the regular staff is being trained. Several training models and/ or methods have been tried:

- Either offer a comprehensive course as an entire day session or offer multiple 1- to 2-hour sessions.

- Offer a short course to cover the absolute minimum amount of functionality and then have specially trained staff available to help physicians learn the remainder of the functions as they care for patients on the wards.

- Develop a self-paced, computer-based training module that the physicians can complete at their leisure.

At the University of Virginia, more than 3600 nurses, 1200 residents, 800 medical students, and 200 attending physicians were trained to use an order-entry system.[7] Physicians were trained in a 6-hour course. Physicians could take the entire course at once or in six 1-hour segments at their convenience. It is not uncommon for individual sign-on codes to be withheld until physicians have completed the training course.[36] It is a common perception that effective physician training requires 2 to 6 hours of one-on-one training, which is best accomplished—or at least tolerated—if presented by another member of the medical staff.[54]

Community Memorial Hospital in Toms River, New Jersey selected a group of 12 nurses to provide training for all 200 physicians at the facility.[50] The trainers were responsible for developing orientation packets and for scheduling training sessions. All training was done in one-on-one sessions lasting 2 hours. During implementation, trainers were available at all nursing units. In addition, demonstrations of the information system were held monthly for physicians.

Training is important, not just to make sure that orders are entered correctly, but also to help physicians learn to be efficient. When Ogura et al. compared two groups of physicians selected by their amounts of computer training, they found that the more experienced physicians entered orders approximately 30% faster (60 seconds versus 88 seconds).[63]

Many investigators have reported that while initially CPOE requires more time to learn than the paper method, over a short period (2 weeks to 2 months), there is no significant difference between paper-based and computer-based order entry times.[64,65] Specifically, Schroeder states that "after 1 or 2 weeks of use, most physicians find they can enter orders in the computer terminals faster than they could write them by hand."[36] One must be careful when assessing these numbers, however, because systems are often customized to meet various requirements of particular institutions. Times to enter orders and times required to learn to use the systems may vary greatly.

1.4.4 Time Required to Enter Orders

In an 11-day study in Japan, investigators found that 5562 prescription orders (2952 new and 2160 repeat orders), 4363 laboratory orders (3820 entered by physicians), and 1218 X-ray orders (1217 entered by physicians) were input to the system.[63] During the study period, the mean time to enter a new prescription was 102 seconds (standard deviation [SD] = 121 seconds), while a laboratory order required 76 seconds (SD = 109 seconds) and a radiologic examination 54 seconds (SD = 82 seconds). The researchers made no timing comparisons with the previous paper-based system.

In a formal time–motion analysis of the effects of using CPOE, Tierney et al. found that physicians spent an average of 33 minutes more writing orders between the hours of 10 A.M. and 8 P.M. than did controls using a paper-based system (58.5 versus 25.5 minutes; $p < 0.001$).[29] This averages out to 5.5 minutes longer per patient writing orders during the study hours. Much of this difference (9.9 of 33 minutes) can be accounted for by the discharge order process. Control physicians simply wrote "discharge patient today," whereas the intervention group had to include prescriptions, discharge planning information, and a brief typed discharge summary. In other words, they took longer but did things that the manual group would have to do outside the order-entry process. A portion of this increase in time was recaptured by the intervention group through a 5.7-minute savings in the time spent managing "scut" cards (short notes that record vital patient information).

In an analysis of the time spent by residents (all services combined) using CPOE, Massaro found that fewer than 10% of the residents spent more than 60 minutes during a 24-hour period on the computer, although on specific high-volume rotations and/or services it was not uncommon for residents to spend 4 to 6 hours at the terminal in a 24-hour period.[7] He also found that first-year residents (who enter the majority of the orders in any system) spent significantly more time at the computer than did the more experienced residents. At Yale–New Haven Hospital, Clyman reported that "a surgery intern entered 475 new orders into the computer in 1 day" and that "only 50 of these originated from order sets" (J. Clyman, 1993, personal communication). These patterns suggest that one person is being assigned the duty of entering orders for a large team. Orders should be entered by the decision maker wherever possible if CPOE is to deliver its full potential benefits.

1.5 System Design Issues

1.5.1 Capabilities/Policies Required by CPOE

CPOE requires features beyond those found in a system designed for clerical users. In most cases, additional policies and system functions must be created for an integrated, efficient system:

- *Electronic signatures*. One must be sure that electronic signatures are legal and binding in the state.[15]

- *Suspended orders*. A mechanism should be in place to allow preadmission orders to be entered for patients before they arrive at the hospital. The orders must remain suspended until the patient is admitted and then automatically activated. A similar mechanism could permit suspension of orders on patient transfer, with selective reactivation and countersignature by the receiving physician.

- *Countersignature orders*. Two types of countersignatures must be supported. The first is legal countersignature of a verbal order. In this case, the order can be activated prior to countersignature but the nurse recording the order must have an electronic signature and must indicate which physician generated the order. The second is countersignature of an order (by a medical student or nonauthorized consultant) that had to be held pending countersignature.[9]

- *Order modifications by medical students*. When an active order is modified, the order is actually discontinued, and a new order is initiated. Provision must be made to hold both actions for countersignature.

- *Consultant orders*. A method must be in place to allow consultants to enter orders. Questions may arise regarding whether these orders should be acted upon without permission from the attending physician.

- *Downtime procedures*. During system downtime, orders are written on a paper order sheet. Who enters these orders once the system comes up if clerks are not used to handling orders? Do physicians need to countersign orders that have been entered into the system by clerks based on the physicians' handwritten orders? Another question that must be answered is this: How are orders that were in the system retrieved and processed once the system goes down?

1.5.2 Data Entry Methods

Experienced users can achieve speeds of up to 15 keystrokes/second (approximately 150 words per minute) on the keyboard, whereas beginners may struggle along at a rate of less than one keystroke per second.[66] Many different hardware modalities have been tried to facilitate data entry, including the light pen,[67] the mouse,[26] the trackball,[68] the touch screen,[55] voice recognition,[69] bar codes,[70] special-purpose (book-style) keyboards,[51] and more recently, the gesture-recognition systems of the pen-based operating systems.[71] Successful systems usually employ one of these modalities in conjunction with the keyboard.

Childs[67] found that light-pen technology along with user-friendly menus allowed users of the TDS system to perform all of their functions with less than 1% typing. Others have experimented—albeit unsuccessfully—with popping up alphanumeric keypads on the screen and allowing users to pick off the numbers or letters that they wish to enter with a mouse.[28] These indirect pointing devices require much more cognitive processing and hand–eye coordination to bring the on-screen cursor to the desired target.[66]

1.5.3 Outpatient Prescription-Writing Systems

Outpatient prescription writing is the area in which the most work has been done regarding tailoring of a user interface for CPOE. Reported average times to generate prescriptions using computer-based CPOE range from 4.2 minutes for complex prescriptions to less than 30 seconds for the simplest prescriptions (similar times were found using paper-based systems).[65,72] The outpatient setting lends itself to CPOE evaluations because only a small number of physicians need to be involved. Prescriptions are a good test case because they represent a large volume of potentially complex orders.

Levit and Garside developed a system to capture prescriptions entered in coded form.[73] Most drug codes utilized the first two letters of the drug name. Following the name of the drug, the physician entered the dosage form. Common abbreviations for this purpose included "T" for tablets or capsules, "S" for solutions, "C" for creams, and so on. Following the dosage form, the unit dose amount was entered. Most of the doses used the first significant digit of the number (e.g., 2 for 250 mg). The physician then entered the code for the number of capsules of the total amount of the drug to be given. Once again, only the first digit of the total amount of the dosage was entered as the code. Thus, to write a simple prescription for 60 250-mg tablets of tetracycline, one would enter TET26. If the code stopped at that point, the directions "Take as directed" would also appear on the label. However, it was preferable, as well as possible, to also enter codes for "how many" and "how often." Once again, the user entered the first digit for "how many" and the first letter of the commonly used Latin abbreviations for "how often" (e.g., 2T for two tablets three times per day—t.i.d.). Defaults for the number of refills could be set or the number allowed could be entered as "R," followed by the number of refills allowed. More complex codes were also possible by including text between asterisks and/or combining instructions with a plus sign. This approach proved efficient for the expert user but is probably not applicable to large physician populations at a medical center because of the extensive training requirements.

Brown et al. developed another innovative user interface to a prescription-writing system that requires only five function keys (the arrow keys) to generate a complete detailed prescription.[74] In addition, medication names are accessed rapidly from a doctor's personalized drug formulary, which is indexed by conventional therapeutic groups and cross-indexed by disease or diagnosis groupings. Finally, extensive use is made of defaults, which allows the physician to order by defaults and exceptions. To order a drug, the physician presses and holds an arrow key that causes a simulated wheel to rotate on the screen (keys are automatically repeating). When the appropriate dose form or timing, or both, is visible, the user presses another key to select that item. Allen et al. later compared order-entry times of novices with those of an experienced user of this system and found that the experienced user required 18 seconds on average to complete the order, whereas the novice users required 82 seconds to do so.[49] This system has been in continuous operation for more than 3 years.

1.5.4 Order Sets

Order sets allow users to issue prepackaged groups of orders applicable to a specific diagnosis or to a specific time period within an episode of care (e.g., admission or postoperative orders). Order sets can reduce the task and error rate of writing common repetitive orders.

Anderson et al.[75] developed a computer-based simulation to represent the process through which physicians and other hospital personnel enter orders into a hospital information system. They found that by increasing the use of order sets by approximately 50%, they could reduce the time hospital personnel spent entering orders by 20% and decrease terminal usage by 30%. In their report on this study, the researchers stated their belief that they could decrease the percentage of undetected errors in orders by 40% using this technique. Order sets allow clinicians to determine appropriate orders away from the time and emotional pressures of the day-to-day clinical setting.[76]

Order sets are usually developed by a group of physicians with a common clinical focus. This process involves a review of current clinical practices followed by development of consensus regarding the best diagnostic and treatment options. Another approach is to "memorize" sets (e.g., save the orders as a by-product of normal order entry and then recall them at a later date) during routine order entry. These so-called personal order sets (POSs) have high physician acceptance. Drawbacks to use of POSs include the fact that they may reinforce inefficient medical care. In addition, the number of POSs may become unmanageable. For example, at the University of Virginia, a resident-led oversight committee was required to reduce the number of POSs generated by 273 residents in the first 2 years of operation from 2684 to 545 (or from approximately 10 per resident to 2 per resident) to improve maintainability of the system.[19] Maintainability becomes a significant problem if either the laboratory or the pharmacy changes a specific test or treatment that is included within multiple order sets. In this case, someone must change every order set or clinicians will have to correct that specific order every time one of the "incorrect" order sets is selected.

Levine et al. state that the benefits associated with order sets can be grouped into two main categories: those that improve the quality of medical care and those that enhance workers' productivity (see Table 1-1).[77] While many clinicians and administrators are strong supporters of the development and use of order sets, others argue that their potential drawbacks limit their utility in many situations. Specifically, concerns have arisen that order sets might lead to the practice of "cookbook" medicine. If they are developed or used indiscriminately, the use of order sets may increase unnecessary orders. Opponents also argue that the use of order sets adversely affects

Table 1-1   Some Benefits of Order Sets

Category	Benefit
Quality	Reduction in transcription errors
Quality	Promotion of adherence to consistent standards of care
Quality	Focus attention on unique features of a patient
Productivity	Quicker order entry
Productivity	Reduction in delays due to inconsistent or incomplete orders

Adapted from: Levine HS, Marino T, Grossman M. *Order sets: Practical issues for implementation. Proceedings of the Healthcare Information and Management Systems Society.* Chicago: American Hospital Association; 1991:313–318.

the educational process, as students do not get the experience of repetitively writing out common orders.

Many types of order sets have been used. The first category lists common orders for a department, service, or patient unit on a menu for rapid selection. Orders are selected from the menus and completed individually. The second category lists detailed orders for a procedure or day of care. Orders are selected as a group with minor editing by exception to reflect patient variations. A third category outlines order options with multiple-choice or fill-in-the-blank fields. These order outlines are a compromise between individual orders, which require many individual steps to complete, and order sets, which offer very little flexibility. A fourth category of order set differs from the first three categories in that it is created on the fly by an algorithm in the clinical information management system that takes into account pertinent clinical data. In the simplest sense, this type of order set may consist of a rank-ordered list of common options.[26] In its most complex sense, the system may suggest a specific order for a particular patient automatically.[47]

In an attempt to improve the acceptance of order sets in general, Anderson et al. conducted a study at Methodist Hospital of Indiana, a large private teaching hospital, to see if they could increase the use of order sets.[78] Their intervention consisted of individual meetings with the physicians identified as being educationally influential. At these meetings, members of the HIS project staff discussed current usage statistics of the HIS in general and order-entry statistics in particular, with emphasis on the advantages of using order sets. The researchers then compared the use of order sets before and after initiation of their intervention. The use of order sets by the physicians in the group headed by the influential physicians significantly increased compared to the control group. The researchers concluded that use of these influential physicians was an effective method of changing physicians' practice patterns with respect to the use of order sets.

1.6 Use of Expert Systems to Facilitate CPOE

One of the primary justifications of CPOE is that computer-generated reminders and/or advice can be provided to the person best able to act on the information at the time and place that the action is required.[79,80] Many such "expert systems" have been developed, but few have been integrated into the normal ordering routine. The following examples illustrate how such systems can work.

1.6.1 Systems to Facilitate Ordering of Radiologic Procedures

The PHOENIX expert system[81,82] is integrated with the Missouri Automated Radiology System (MARS) computer system, a dedicated radiology information management system[83] used at the University of Chicago Medical Center. The prototype system contains information for 10 common procedures (e.g., chest, abdomen, and cervical spine radiography; abdominal computed tomography; gallbladder sonography). PHOENIX can assist the ordering physician by providing reminders, such as asking whether females between the ages of 6 and 60 might be pregnant, before completing the order for a usual two-view chest radiograph. PHOENIX reminds the physician that it may be inappropriate to perform a contrast-enhanced CT on a patient with an elevated serum creatinine level unless certain conditions are met (e.g., dialysis is provided following the procedure). In addition, PHOENIX interviews the ordering physician to help determine whether a screening mammogram (appropriate for an asymptomatic woman with no prior breast abnormality) or a dedicated mammogram (which requires physician supervision and possible additional views) is indicated.

DxCON, developed by researchers at Yale University as a prototype artificial intelligence-based computer system, gives advice to physicians regarding the optimal sequencing of radiologic tests for diagnosis of obstructive jaundice.[84] This systems utilizes a "critiquing mode" of interaction, in which the computer does not tell the physician what to do, but rather asks the physician about the patient and the intended workup plan and then discusses the strengths and weaknesses of that plan in an English-prose discussion that is tailored to the physician's specific plan. In this way, the physician can evaluate the appropriateness of the computer's conclusions and advice regarding a specific patient. In several example cases, DxCON has produced convincing critiques of several different workup plans.[84]

1.6.2 Systems to Facilitate Transfusion Orders

Using HELP, Lepage et al. developed a consultation module that recommended the type of blood product (red blood cells or platelets) to be used, authorized the number of units to be ordered, and suggested an ordering priority.[48] In a 3-month retrospective study, they found that the computer-based blood-ordering consultant accurately recommended 95.5% of the orders (for which data were available in the computer) that were entered. These researchers also evaluated the quantities of blood products ordered and found agreement in 71.2% of the cases. In the remainder of the cases, the computer recommended smaller numbers of units to be ordered. Lepage et al. concluded that their consultation system could simplify the blood-ordering process and reduce the number of units ordered. A clinical trial comparing the critiquing mode with the consultation mode is planned.

ESPRE, a knowledge-based expert system developed at the University of Minnesota Hospital and Clinic, provides automated decision support for blood bank personnel in assessing requests for platelets.[85] ESPRE uses a hybrid rule-based and frame-based knowledge-base structure to provide information about diagnoses and conditions such as disseminated intravascular coagulation, prolonged coagulation time, poor platelet activation, infections, and surgical procedures that affect transfusion requests. In a random sample of 75 platelet transfusion requests, ESPRE's recommendations agreed with those of blood bank personnel 93% of the time.[85]

In a similar vein, Spackman et al.[86] developed the Transfusion Advisor (TA), which critiques the appropriateness of orders for cryoprecipitate, frozen plasma, and platelets. In a review of 31 consecutive requests for these blood products, there was total agreement between the TA and the medical director of the blood bank in favor of carrying out the order in nine cases. In an additional 19 cases, both the medical director and the TA decided that transfusions were not required. In three cases, the TA's knowledge base was found to be deficient and promptly fixed. The authors noted that if the TA had been responsible for dispensing the blood products that were ordered in these 31 cases, the hospital could have saved more than $2200 (60% reduction in costs).

1.7 Implications for Future CPOE Systems

Past experience suggests the following ingredients are key to successful implementation of CPOE.

First, the system must be fast (sub-second response time) and must be easy to use with minimal training. In the event that assistance is required, it should be available 24 hours a day, both online and by telephone. Consistency of the system interface and behavior may be more important than having different screens and/or responses tailored to specific situations.

Second, broad and committed involvement and direction by physicians prior to implementation is vital. CPOE must have real and committed sponsorship within the clinical community.

Third, the top leadership of the organization must be committed to stay the course in terms of implementation of the system. Problems will inevitably occur; they will be solved only if people focus on making the system work better instead of being diverted by the question of whether the institution should be implementing CPOE.

Finally, a group of problem solvers must meet regularly with users to work out procedural issues that cross boundaries. This group must include someone from each of the following areas: attending physicians, house staff, nursing, admitting, laboratory, pharmacy, radiology, billing, and information management. The group members must also be empowered to commit to specific decisions that require changes in the areas they represent. The goal should be to discuss a problem and agree on a solution one week and to implement that solution prior to the next week's meeting.

Optimistic forecasts of widespread, rapid adoption of CPOE must be balanced by three observations:

- The proven systems that support CPOE were designed in the 1970s. A decision must be made to install an old "tried-and-true" system or to be part of a new experiment. Either course has risks.
- CPOE requires change in the way healthcare professionals work. To have a good outcome, they must put their time into the implementation effort not just at the beginning of the project but in an ongoing fashion.
- CPOE must be part of a comprehensive clinical information management environment. Physicians readily appreciate the advantage of clinical data retrieval, and "smart" CPOE systems need clinical data to work.

Physician order entry may well come into common use in the near future. The potential benefits of CPOE are compelling. Health care is one of the last major industries to rely on pen and paper for the majority of its record keeping. Healthcare reform, however, promises to bring changes in clinical practice and teaching patterns—and a shift to CPOE can be incorporated into those changes. With proper planning, CPOE may lessen the impact of these changes by incorporating state-of-the-art information management into the physician's work patterns.

1.8 References

1. Collen MF. General requirements for a medical information system (MIS). *Comput Biomed Res*. 1970;3:393–406.
2. Buchanan NS. Evolution of a hospital information system. *Proc Annu Symp Comput Appl Med Care, IEEE*. 1980;14:34–36.
3. Fischer PJ, Stratmann WC, Lundsgaarde HP, Steele DJ. User reaction to PROMIS: Issues to acceptability of medical innovations. *Proc Annu Symp Comput Appl Med Care, IEEE*. 1980;4:1722–1730.
4. Dambro MR, Weiss BD, McClure CL, Vuturo AF. An unsuccessful experience with computerized medical records in an academic medical center. *J Med Educ*. 1988;63:617–623.
5. Hodge MH. Direct use by physicians of the TDS medical information system. In: Blum BI, Duncan K, Eds. *A History of Medical Informatics*. New York: ACM Press; 1990:345–356.
6. Williams LS. Microchips versus stethoscopes: Calgary hospital, MDs face off over controversial computer system. *Can Med Assoc J*. 1992;147:1535–1547.
7. Massaro TA. Introducing physician order entry at a major academic medical center: I. Impact on organizational culture and behavior. *Acad Med*. 1993;68:20–25.
8. Dorenfest SI. History and impediments to progress in the development and implementation of the computerized patient record. *Proc Healthcare Info Manage Syst Soc*. 1993;2:81–92.
9. Teich JM, Hurley JF, Beckley RF, Aranow M. Design of an easy-to-use physician order-entry system with support for nursing and ancillary departments. *Proc Annu Comput Appl Med Care*. 1993;16:99–103.
10. Stead WW, Borden R, McNulty P, Sittig DF. Building an information management infrastructure in the 90s: the Vanderbilt experiment. *Proc Annu Symp Comput Appl Med Care*. 1994;17:534–538.
11. Weiser M. The computer for the 21st century. *Sci Am*. 1991;265:94–104.
12. Dasta JF, Greer ML, Speedie SM. Computers in healthcare: overview and bibliography. *Ann Pharmacother*. 1992;26:109–117.
13. Haynes RB, McKibbon KA, Bayley E, Walker CJ, Johnston ME. Increases in knowledge and use of information technology by entering medical students at McMaster University in successive annual surveys. *Proc Annu Symp Comput Appl Med Care*. 1993;16:560–563.
14. Dick RS, Steen E. The computer-based patient record: an essential technology of health care. Washington, DC: Institute of Medicine; 1991.
15. Schoenbaum SC, Barnett GO. Automated ambulatory medical records systems: an orphan technology. *Int J Tech Assess Health Care*. 1992;8:598–609.

16. Barrett JP, Barnum RA, Gordon BB, Pesut RN. *Final report on evaluation of the implementation of a medical information system in a general community hospital.* Battelle Laboratories (NTIS PB 248 340); December 19, 1975.

17. Barrett JP, Hersch PL, Caswell RJ. *Evaluation of the impact of the implementation of the Technicon Medical Information System at El Camino Hospital: Part II. Economic trend analysis.* Battelle Columbus Laboratories (NTIS PB 300 869); May 14, 1979.

18. Grams S, Grieco A, Williams R. *Physician use of the TDS medical information system provides major benefits and savings to New York University Medical Center.* TDS Technical Report, 1986.

19. Massaro TA. Introducing physician order entry at a major academic medical center: II. Impact on medical education. *Acad Med.* 1993;68:25–30.

20. Gardner RM, Golubjatnikov OK, Laub RM, Jacobson JT, Evans RS. Computer-critiqued blood ordering using the HELP system. *Comput Biomed Res* 1990;23:514–528.

21. Warner HR, Olmsted CM, Rutherford BD. HELP: A program for medical decision making. *Comput Biomed Res.* 1972;5:65–74.

22. Pryor TA, Gardner RM, Clayton I'D, Warner HR. The HELP system. *J Med Sys.* 1983;7:87–101.

23. Kuperman GJ, Gardner RM, Pryor TA. HELP: *A Dynamic Hospital Information System.* New York: Springer-Verlag; 1991.

24. Lepage EF, Gardner RM, Laub RM, Golubjatnikov OK. Improving blood transfusion practice: role of a computerized hospital information system. *Transfusion.* 1992;32:253–259.

25. McDonald CJ. Protocol-based computer reminders, the quality of care and the nonperfectibility of man. *N Engl J Med.* 1976;295:1351–1355.

26. McDonald CJ, Tierney WM. The medical gopher: a microcomputer system to help find, organize and decide about patient data. *West J Med.* 1986;145: 823–829.

27. McDonald CJ. *Action-Oriented Decisions in Ambulatory Medicine.* Chicago: Year Book Medical Publishers; 1981.

28. McDonald CJ, Tierney WM, Martin DK, Overhage JM, Day Z. The Regenstrief medical record: 1991. A campus-wide system. *Proc Annu Symp Comput Appl Med Care.* 1991;15:925–928.

29. Tierney WM, Miller ME, Overhage JM, McDonald CJ. Physician inpatient order writing on microcomputer workstations: effects on resource utilization. *JAMA.* 1993;269:379–383.

30. Glaser JP, Teich J, Kuperman G. The future of clinical information systems: One hospital's perspective. *Top Health Inform Manage.* 1993;14:12–24.

31. Tierney WM, Miller ME, McDonald CJ. The effect on test ordering of informing physicians of the charges for outpatient diagnostic tests. *N Engl J Med.* 1990;322:1499–1504.

32. Ogura H, Sagara E, Iwata M, et al. Online support functions of prescription order system and prescription audit in an integrated hospital information system. *Med Inform.* 1988;13:161–169.

33. Evans RS, Pestotnik SL, Burke JP, et al. Reducing the duration of prophylactic antibiotic use through computer monitoring of surgical patients. *DICP Ann Pharmacother.* 1990;24:351–354.
34. Ogura H, Yamamoto K, Furutani H, Kitazoe Y, Hirakawa Y, Sagara E. Online prescription order and prescription support in an integrated hospital information system. *Med Inform.* 1985;10:287–299.
35. Hodge MH. History of the TDS medical information system. In: Blum BI, Duncan K, Eds. *A History of Medical Informatics.* New York: ACM Press; 1990:328–344.
36. Schroeder CG, Pierpaoli PG. Direct order entry by physicians in a computerized hospital information system. *Am J Hosp Pharm.* 1986;43:355–359.
37. Reynolds MS, Kirkwood CF, Ostrosky JD, Cessna LD, Clapham CE. *Effect of an educational computer screen on direct physician order entry of anti-anaerobic drugs.* 23rd Annu Am Soc Hosp Pharm (ASHP) Mid-year Clinical Meeting, Dallas, TX, 1988.
38. Kawahara NE, Jordan FM. Influencing prescribing behavior by adapting computerized order-entry pathways. *Am J Hosp Pharm.* 1989;46:1798–1801.
39. Perry M, Myers CE. *Computer cost prompting as a determinant of hospital drug prescribing* (abstr). 18th Annu Am Soc Hosp Pharm (ASHP) Mid-year Clinical Meeting, Atlanta, GA, 1983.
40. Tierney WM, McDonald CJ, Martin DK, Rogers MP. Computerized display of past test results: effect on outpatient testing. *Ann Intern Med.* 1987;107:569–574.
41. Tierney WM, McDonald CJ, Hui SL, Martin DK. Computer predictions of abnormal test results: effects on outpatient testing. *JAMA.* 1988;259:1194–1198.
42. East T, Morris AH, Wallace CJ, et al. A strategy for development of computerized critical care decision support systems. *Int J Clin Monit Comput.* 1991;8:263–269.
43. Picart D, Guillois B, Nevo L, Alix D. A program for parenteral and combined parenteral and enteral nutrition of neonates and children in an intensive care unit. *Intens Care Med.* 1989;15:279–282.
44. Halpern NA, Thompson RE, Greenstein RJ. A computerized intensive care unit order-writing protocol. *Ann Pharmacother.* 1992;26:251–254.
45. Goldberg, DB, Baardsgaard G, Johnson MT, Jolowsky CM, Sheperd M, Peterson CD. Computer-based program for identifying medication orders requiring dosage modification based on renal function. *Am J Hosp Pharm.* 1991;48:1965–1969.
46. White KS, Lindsay A, Pryor TA, Brown WF, Walsh K. Application of a computerized medical decision-making process to the problem of digoxin intoxication. *J Am Coll Cardiol.* 1984;4:571–576.
47. Sittig DF, Pace NL, Gardner RM, Beck E, Morris AH. Implementation of a computerized patient advice system using the HELP clinical information system. *Comput Biomed Res.* 1989;22:474–487.
48. Lepage EF, Gardner RM, Laub RM, Jacobson JT. Assessing the effectiveness of a computerized blood order "consultation" system. *Proc Annu Symp Comput Appl Med Care.* 1992;15:33–37.
49. Allen SI, Johannes RS, Brown CS, Kafonek DM, Plexico PS. Prescription-writing with a PC. *Comput Meth Progr Biomed.* 1986;22:127–135.
50. Larson RL, Blake JP. Achieving order entry by physicians in a computerized medical record. *Hosp Pharm.* 1988;23:551–553.

51. Yokoyama J, Fukuda A. *Prescription order service in Kanto Teishin hospital information system.* MEDINFO 80. Amsterdam, Netherlands: IFIP, North Holland; 1980:929–934.

52. Morris AH. Evaluation of new therapy: extracorporeal CO, removal, protocol control of intensive care unit care, and the human laboratory. *J Crit Care.* 1992;7:280–286.

53. Blackmon PW, Marino CA, Aukward RK, et al. *Evaluation of the medical information system at the NIH clinical center.* Analytic Services, Inc. (NTIS PB 82–190083); 1982.

54. Bria WF, Rydell RL. *The Physician–Computer Connection.* Chicago: American Hospital Publishing; 1992.

55. Spillane MJ, McLaughlin MB, Ellis KK, Montgomery WL, Dziuban S. Direct physician order entry and integration: potential pitfalls. *Proc Symp Comput Appl Med Care.* 1990;14:774–778.

56. Ellinoy BJ. Fax fiction. *Am J Hosp Pharm.* 1989;46:1549–1550.

57. McAllister JC. Pharmacy fax. *Am J Hosp Charm.* 1989;46:255–256.

58. Tierney WM, Overhage JM, McDonald CJ, Wolinsky FD. Medical student and housestaff opinions of computerized orderwriting. *Acad Med.* 1994;69: 386–389.

59. Haynes RB, McKibbon KA, Walker CJ, Ryan N, Fitzgerald D, Ramsden MF. Online access to MEDLINE in clinical settings: a study of use and usefulness. *Ann Intern Med.* 1990;112:78–84.

60. Kingsland LC, Harbourt AM, Syed EJ, Schuyler PL. COACH: Applying UMLS knowledge sources in an expert searcher environment. *Bull Med Libr Assoc.* 1993;81:178–183.

61. Barnett GO, Cimino JJ, Hupp JA, Hoffer EP. DXplain: an evolving diagnostic decision-support system. *JAMA.* 1987;258:67–74.

62. Miller RA, Massarie FE, Myers JD. Quick Medical Reference (QMR) for diagnostic assistance. *MD Comput.* 1986;3:34–48.

63. Ogura H, Sagara E, Yamamoto K, Furutani H, Kitazoe Y, Takeda Y. Analysis of the online entry process in an integrated hospital information system. *Comput Biol Med.* 1985;10:287–299.

64. Reynolds RE, Heller EE. An academic medical center experience with a computerized hospital information system: the first four years. *Proc Annu Symp Comput Appl Med Care, IEEE.* 1980;4:3–16.

65. Garrett LE, Hammond WE, Stead WW. The effects of computerized medical records on provider efficiency and quality of care. *Meth Inform Med.* 1986;25:151–157.

66. Shneiderman B. *Designing the User Interface: Strategies for Effective Human–Computer Interaction.* Reading, MA: Addison-Wesley; 1987.

67. Childs BW. El Camino/National Institutes of Health: A case study. In: Bakker AR, Ball MJ, Scherrer JR, Willems JL, Eds. *Towards New Hospital Information Systems.* Amsterdam, Netherlands: Elsevier Science Publishers (North Holland); 1988:83–89.

68. Carr D, Hasegawa H, Lemmon D, Plaisant C. The effects of time delays on a telepathology user interface. *Proc Annu Symp Comput Appl Med Care.* 1993;16:256–260.

69. Klatt EC. Voice-activated dictation for autopsy pathology. *Comput Biol Med.* 1991;21:429–433.

70. Gouveia WA. Managing pharmacy information systems. *Am J Hosp Pharm.* 1993;50:113–116.

71. Lussier YA, Maksud M, Desruisseaux B, Yale PP, St-Arneault R. PureMD: a computerized patient record software for direct data entry by physicians using a keyboard-free pen-based portable computer. *Proc Annu Symp Comput Appl Med Care.* 1993;16:261–264.

72. Donald JB. Online prescribing by computer. *Br Med J.* 1986;292:937–939.

73. Levit F, Garside DB. Computer-assisted prescription writing. *Comput Biomed Res.* 1977;10:501–510.

74. Brown CS, Allen SI, Songco DC. A computerized prescription writing program for doctors. *Meth Inform Med.* 1985;24:101–105.

75. Anderson JG, Jay SJ, Clevenger SJ, Kassing DR, Perry J, Anderson MM. Physician utilization of a hospital information system: a computer simulation model. *Proc Annu Symp Comput Appl Med Care, IEEE.* 1988;12:858–861.

76. Sittig DF, Gardner RM, Morris AH, Wallace CJ. Clinical evaluation of computer-based respiratory care algorithms. *Int J Clin Monit Comput.* 1990;7:177–185.

77. Levine HS, Marino T, Grossman M. Order sets: practical issues for implementation. *Proc Healthcare Info Manage Syst Soc.* 1991:313–318.

78. Anderson JG, Jay SJ, Perry J, Anderson MM, Schweer HM. Informal communication networks and change in physicians' practice behavior. *Proc Conf Res Med Educ.* 1988;27:127–132.

79. Greenes RA, Tarabar DB, Krauss M, et al. Knowledge management as a decision support method: a diagnostic workup strategy application. *Comput Biomed Res.* 1989;22:113–135.

80. Sanders GD, Lyons EA. The potential use of expert systems to enable physicians to order more cost-effective diagnostic imaging examinations. *J Dig Imag.* 1991;4:112–122.

81. Kahn CE, Messersmith RN, Jokich MD. PHOENIX: An expert system for selecting diagnostic imaging procedures. *Invest Radiol.* 1987;22:978–980.

82. Kahn CE, Kovatsis PC, Messersmith RN, Lehr JL. Automated entry of radiology requisition information with artificial intelligence techniques. *Am J Roentgenol.* 1989;153:1085–1088.

83. Lehr JL, Lodwick GS, Nicholson BF, Birznieks FB. Experience with MARS (Missouri Automated Radiology System). *Radiology.* 1973;106:289–294.

84. Swett HA, Rothchild M, Weltin GG, Fisher PR, Miller PL. Optimizing radiologic workup: an artificial intelligence approach. *J Dig Imag.* 1989;2:15–20.

85. Sielaff BH, Connelly DP, Scott EP. ESPRE: A knowledge-based system to support platelet transfusion decisions. *IEEE Trans Biomed Eng.* 1989;36:541–546.

86. Spackman KA, Chabot MJ, Beck JR. The transfusion advisor: a knowledge-based system for the blood bank. *Proc Annu Symp Comput Appl Med Care, IEEE.* 1988;12:18–21.

2   An Introduction to Unintended Consequences of Clinical Information Technology

Based on: Campbell E, Sittig DF, Ash JS, Guappone K, Dykstra R. Types of unintended consequences related to computerized provider order entry. *J Am Med Inform Assoc.* 2006;13(5):547–556.

Ash JS, Sittig DF, Poon EG, Guappone K, Campbell E, Dykstra RH. The extent and importance of unintended consequences related to computerized provider order entry. *J Am Med Inform Assoc.* 2007;14(4):415–423.

2.1 Key Points

- Computerized physician order-entry (CPOE) systems can help hospitals reduce medical errors, but they can also introduce new risks.

- The unintended consequences of CPOE are widespread and important to those knowledgeable about CPOE in hospitals. They can be positive, negative, or both, depending on one's perspective, and they continue to exist over the duration of use.

- Unintended adverse consequences associated with CPOE can be classified into nine major categories (in order of decreasing frequency): (1) more/new work for clinicians; (2) unfavorable workflow issues; (3) never-ending system demands; (4) problems related to paper persistence; (5) untoward changes in communication patterns and practices; (6) negative emotions; (7) generation of new kinds of errors; (8) unexpected changes in the power structure; and (9) overdependence on the technology.

- Identifying and understanding the types, and in some instances the causes, of unintended adverse consequences associated with CPOE will enable system developers and implementers to better manage implementation and maintenance of future CPOE projects. Aggressive management of unintended adverse consequences is vital for system success.

2.2 Introduction

Healthcare organizations often implement CPOE as part of their approach to improve medication safety and reduce healthcare costs.[1–3] Yet several studies indicate that unpredictable, emergent problems, or *unintended adverse consequences* (UACs) can surround CPOE implementation and maintenance.[4,5] Careful identification, description, and categorization of UACs can provide insight into the unexpected outcomes of placing CPOE systems into complex healthcare work environments.

The purpose of this book is to identify and describe the major types of UACs related to CPOE implementation. Because CPOE implementations affect many different types of personnel in the healthcare environment, their evaluation must encompass multiple, discrete perspectives. The work described here focuses *not* on the impact of CPOE on clinical outcomes for patients, but rather on its impacts on the healthcare personnel who use, maintain, or manage CPOE systems. Specifically, the perspectives of three groups are considered: clinical end users, IT staff, and clinical administrators. We broadly define *clinical end users* as those healthcare providers and other clinical staff (e.g., physicians, pharmacists, nurses, ward secretaries) who work with CPOE systems. *IT staff* includes those individuals who implement, configure, maintain, and support CPOE systems, whether their primary professional background is technical or clinical in nature. Finally, *administrative staff* refers to those persons who manage the organizational implementation of CPOE, through establishing policies and procedures, assuring compliance with local and federal guidelines, and making high-level CPOE-related resource allocation decisions.

2.3 Background

2.3.1 Theoretical Framework

Diffusion of innovations (DOI) theory served as the framework for the study described in this book. *Diffusion* has been defined by Everett Rogers as "the process by which an innovation is communicated through certain channels over time among the members of a social system," whereas an *innovation* is "an idea, practice, or objective perceived as new by an individual, a group, or an organization."[6] While adoption of any innovation inevitably generates consequences, Rogers notes that such consequences constitute the least studied aspect of diffusion of innovation. DOI theory suggests that consequences can be desirable or undesirable, and anticipated or unanticipated. In this study, *unintended consequences* refers to events that are neither anticipated nor the specific goals of the associated CPOE project. Although "unintended" most often connotes consequences that are both unanticipated and undesirable, we have found, in our work, numerous unintended, desirable, positive, beneficial consequences of CPOE as well.[7] While this book focuses on unintended adverse consequences of CPOE adoption for the most part, it is important to remember that unintended consequences are not uniformly errors or mistakes: They are simply surprises that can span a spectrum from lucky to unfortunate. Errors and adverse events comprise a subset of all consequences.

2.4 Major Categories of Unintended Consequences Identified

The study described in this book took place at the sites identified in Table 2–1. Nine major types of unintended consequences emerged from the data. Table 2-2 lists the UACs and their frequencies of occurrence. A detailed discussion of each UAC type

Table 2-1   Description of Sites Studied

Hospital	Size (Number of Beds)	Type of Institution	CPOE System	CPOE Implementation Date	Percentage of Orders Entered via CPOE
Wishard Memorial, Indianapolis, IN	340	Acute care county teaching hospital associated with Indiana University School of Medicine	Homegrown: Regenstrief Medical Records System (RMRS)	1973	100%
Massachusetts General Hospital, Boston, MA	893	Large, academic, general hospital; part of Partners HealthCare System; associated with Harvard Medical School	Homegrown: Clinical Application Suite	1994	100%
The Faulkner, Boston, MA	150	Community teaching hospital with a private medical staff, affiliated with Harvard Medical School	Meditech	2003	95%
Brigham and Women's Hospital, Boston, MA	725	Large, academic, general hospital; part of Partners HealthCare System; associated with Harvard Medical School	Homegrown: BICS	1991	90%
Alamance Regional Medical Center, Burlington, NC	238	Community and teaching hospital	Eclipsys	1998	95%

Table 2-2   Unintended Consequences of CPOE Implementation and Their Frequencies of Occurrence

Unintended Consequence	Frequency (%) (n = 324)
More/new work for clinicians	19.8
Workflow issues	17.6
Never-ending system demands	14.8
Paper persistence	10.8
Changes in communication patterns and practices	10.1
Emotions	7.7
New kinds of errors	7.1
Changes in the power structure	6.8
Overdependence on technology	5.2
Total	**100**

follows, including direct quotations from speakers who articulated an issue particularly well. Table 2-3 includes additional quotes relevant to each type of UAC. Because study subjects were promised confidentiality, researchers edited statements to project confidentiality whenever original statements potentially identified either speakers or the observation site.

Following this introductory chapter, the remaining chapters in Part 1 of the book describe each of these UACs in much greater detail.

2.4.1 Type 1: More/New Work for Clinicians

Clinical information systems can potentially create new work for all staff members (e.g., both clinical and nonclinical staff). The first type of UAC focuses on the ever-increasing workload of clinicians. Despite the common CPOE implementation goal of providing a better "patient overview" to the clinician, many CPOE systems make clinicians do more work to get this overview than before CPOE implementation. The CPOE systems may engender new work by requiring that clinicians perform the following tasks:

- Enter new information (e.g., justification for a treatment selection) not previously required
- Respond to excessive alerts that may contain nonhelpful information (e.g., non-specific medication interactions with no application to the current patient)
- Expend extra time in completing nonroutine, complex orders (e.g., selecting among differing doses and types of insulin to be administered at different times for a diabetic patient)

Many CPOE systems slow the speed at which clinicians can carry out clinical documentation and ordering processes.[8] This loss of efficiency often diminishes over

Table 2-3   Study Participants' Comments on Each Category of Unintended Consequences

Category	Quote
More work/new work	Greater than 95% of the alerts that were generated to physicians turned out to be inconsequential to the patient, and that's a problem.
	Sometimes you get the software and . . . it actually doesn't do what you wanted it to do and the physicians ended up being basically the beta testers.
	Illegibility is not an issue with CPOE, but a lot of time is spent deciphering illogical orders and deciding if that was what the doctor meant or not.
	The junior resident has just inadvertently dragged his cuff across the keyboard while moving from the mouse to the keyboard. This has brought up a login screen of some sort so that the user can cancel some sign-off screen. "This happens all of the time," he says. "Now I am going to have to cancel the order and start again."
Workflow	Because patients are going from the screening room to the pod room to a labor room to the delivery room to postpartum, and each of those is a different level of care, . . . orders need to be rewritten and written . . . the computer is trying to execute JCAHO rules about changing orders for every level of care [and it's] not nearly so flexible . . . Some orders [are] written by certain specialists like the anesthesiologist [for] epidurals. No one wants to rewrite those orders. So how would those [orders] traverse the levels of care when the epidural catheter moves with the patient?
	There is a major problem with confusion over whether the floor is accepting the patient or the ER [is] transferring. The difference is who is responsible for getting things to happen. If orders are entered as the patient goes up, if it is before the ER D/C, then orders print in the ER and not on the floor.
	Unfortunately, now the alerts come up with the discharge summary (e.g., when the patient is about to leave) and they aren't very effective.
	When patients come in [to the ED], they're registered over here . . . Now the patient is in the back, being cared for, but until he or she is registered, there's no way the doctor can enter any of the orders . . . so you have to call X-ray or EKG, because you can't use the system.
Never-ending system demands	You sort of feel you trained people to use the system and then they go live with the system . . . and that's it, right? But it's not. The training needs to be continuous, because not only do they not learn all the things that they really need to learn to make the system efficient initially, . . . but the system changes over time and they always can learn more in terms of training. So training becomes an ongoing, continuous thing.

Continued

Table 2-3   Study Participants' Comments on Each Category of Unintended Consequences (*Continued*)

Category	Quote
	[CPOE] is becoming a battleground for capturing the eyeballs of doctors in terms of ordering the right drug from the pharmacy. The benefits manager wants the cheapest drug. Now pharmaceutical companies are looking to put ads on EMR systems . . . There is also a battle now with the publishing companies to get their products to become the records materials that are linked with the CPOE system.
	They do not have a spell checker and their synonym list was poor, and to resolve the problem, they had to collect every order for allergies over the last one and a half years (about 97,000 of them) and evaluate them. Among other things, they found that 30,000 orders contained some variant of "NKA" or "None known."
	We seem to be doing more things that are technical rather than clinical now.
Paper persistence	When the resident prints out a copy of the patient's electronic medical record, there is a note at the bottom of the page that says "Not intended for the medical record." The residents routinely black out this message, write their additional findings, sign the paper, and [place it in] the chart as progress notes.
	When we moved toward getting them the capability [of electronic order signing,] it made a big difference down here [in Medical Records] because it meant less paperwork. We didn't have a computer system to enter deficiencies in at the time; it all had to be done manually, and it was a lot of labor-intensive-type work moving those charts around.
	They still document their care on flow sheets, med sheets, and the patient chart, but the computer is the source for current orders and lab results.
	The computer-generated documentation seems to take up more pages of a record than the handwritten documents used to.
Communication	When you have a paper order sheet on the unit where the patient [is], that forces the provider to come to the unit. [But by] allowing [providers] to enter orders remotely . . . we found in some cases that you did potentially miss opportunities while you were there to discuss discharge planning or some other kind of clinical problem . . . other types of communication were disrupted by that . . . virtual encounter instead of a real live person being there.
	Communication has helped break down the walls between disciplines. They have to communicate with each other with CPOE. Pharmacy and nursing do very well in this respect.

Table 2-3   Study Participants' Comments on Each Category of Unintended Consequences (*Continued*)

Category	Quote
	In pharmacy, . . . I know the orders kind of go through a pharmacist who will call for clarification if there is any question. Fortunately, with the physician order entry there is less need for clarification because of the way the screen is going to be set up. It collects the dose, the time, and those essential things for a medication order that you don't always get when you're relying on handwritten orders. So a computerized order entry can structure the order set to make sure that the necessary information has been provided at the time of the order, in the majority of cases.
Emotions	At first I hated every second of it. I mean, we were all like, "I have sick patients here. I'm busy. I don't have time to sit here for 20 minutes." It was a pain.
	Now I just wish they would tear up every paper order sheet and process note for doctors so they would have to use the system!
	Well, reams of paper were coming out of the printer in the ER, and the surgeon . . . came in with a bundle of paper and just walked over to where we were working and threw it on the [system trainer] and said, . . . "We're not going to have all this paper! This is just ridiculous!" And he just throws it on her and he just walks out. And we were all stunned.
New kinds of errors	I've noticed the wrong paperwork in the wrong [paper] chart. You just have to tear it out. I don't think we've made it any better since now we've got all the patients in a drop-down [list], with the implication it is easy to pick the wrong one.
	He had ordered an ultrasound but changed his mind to a CT before finalizing the ultrasound entry. He canceled the ultrasound, but somehow it was ordered anyway, and the patient got both studies.
	A patient was in the hospital to rule out MI. A final troponin value was needed to definitely exclude an MI prior to getting an exercise treadmill test [ETT]. If the ETT was negative, the patient could go home. The final lab test was scheduled to be drawn at 1 P.M., but that would be too late to have the results back in time. [The doctor] changed the blood draw to 11 A.M. This was incompatible with the system somehow, and the lab was canceled when the order was entered. [The doctor] was never notified, the lab test never happened, so neither did the ETT, and the patient had to stay in the hospital overnight.
	The HICMA CMS codes that allow you do to things with updating . . . , for fraud and stuff like that, change periodically—but it doesn't change in the system. The codes for doing things are still the old codes, and you can all of a sudden become fraudulent, even though you're doing the right thing.

Continued

Table 2-3   Study Participants' Comments on Each Category of Unintended Consequences (*Continued*)

Category	Quote
Changes in the power structure	It can be about the mismatch of those in charge of implementing the system and those using it because an administrator could think that it's good . . . to send out 20 alerts if there's one that's going to be right on target, but if you ask the physicians to vote on that, and if they're not . . . employed by the institution, they're not going to vote on a 20-to-1 ratio. They're not going to approve that.
	Nurses do enter orders for doctors, most commonly when the doctor calls something in, but most nurses are good at encouraging doctors to enter their own orders.
	When a physician has incomplete records that are over the 14-day regulation, he gets a letter generated every week saying, "You have records [to sign, so] come in." If he doesn't come in and complete those records, his privileges are suspended until they're done, because we have very tight constraints with the state that we have to get those records completed within 14 days.
	When the patient is discharged, all of the paperwork comes down [to Medical Records]. A couple of months ago, we started monitoring for . . . unsigned orders . . . We had actually printed out . . . lists that went to the department chiefs of the doctors that were not signing their orders, and so the physicians had more control of their partners at that time.
Overdependence on technology	We have a resident this year . . . who was voted the best resident by the medical students 2 years in a row, received resident of the year award from the faculty when he completed the program . . ., a wonderful guy. And [he] took a position as a hospitalist at [a hospital] . . . [T]he head of the medical staff called the residency program director about a month after he got there and said, "I just don't understand, this guy is nonfunctional. How [could] you give him such a glowing recommendation?" . . . [The resident] had trained in an environment where he was [using order entry] . . . [He] did not know how to work in a place that didn't have order entry or results retrieval . . . he took almost 6 months to re-acclimate, [to] figure out how to order in a different environment.
	The difference between the medication order list and the medication administration list causes the physicians to have a false sense of security. Specifically, many physicians assume that all of the medications that have been ordered have been administered.
	They use a whiteboard screen saver in the ER that keeps track of people in the ER. When the system goes down or the registry can't supply the ID number, it wreaks havoc in the ER. They don't have a paper backup system of the ER.

time, however.[9] Simply learning to use CPOE takes time and attention away from providers' demanding schedules. If their patient loads are not decreased temporarily during training periods, clinicians must work longer hours to complete their combined electronic and clinical work.[10] The indiscriminant, excessive generation of clinical alerts by CPOE systems can also slow clinicians as they pause to decipher alerts, deliberate on whether and how to respond, and potentially document reasons for not complying with alerts.

Administrators and researchers commonly leverage CPOE to collect information not directly related to patient care. The time burden for doing so usually falls on clinicians, as one study participant noted:

> It seems like every new organizational mandate filters down to the . . . fingertips . . . of our primary care physicians in the form of something else that needs to be entered through the computer and the feeling is, "Well, they have a computer, so it's easy for them to do that," but the cumulative effect [on the physicians] of all those tasks is not fully appreciated.

When CPOE systems are poorly integrated with other clinical information systems, clinicians find it time-consuming to log in to different systems using different account names and passwords. In some cases, data from one system must be entered manually into another system, doubling the work. In addition, built-in functionality such as "cut and paste" features may proliferate redundant text in electronic records that clinicians must navigate to obtain a complete picture of the patient:

> There is no way for me to really know what's new, but I keep seeing chunks of the same text over and over so I have to read every word. Most of it isn't useful.

2.4.2 Type 2: Unfavorable Workflow Issues

Clinical information systems (CIS) in general, and some CPOE systems in particular, by rigidly modeling work processes according to the "letter of the law" (as set forth in organizational policy and procedure) can dramatically highlight mismatches between *intended* and *actual* work processes in real-world clinical settings. These systems can add to ineffective or dysfunctional workflows when CPOE developers' computational models do not reflect actual clinical practices. Such failures can shed light on previously misunderstood blurring across clinical role boundaries (due to confusion, misunderstanding, or duplication), and uncover misconceptions among clinical team members regarding what specific work processes actually entail. Furthermore, if CPOE designers have not considered the appropriate range of workflow perspectives (e.g., those of the nursing or clerical staff as well as those of physicians), the resulting technological system cannot accommodate comprehensive, fully integrated clinical workflows.

When the system fails to support all of the individual role-players who must interact with it, work may be shifted to others, leading to resentment and ineffective work activity synchronization. The following quote illustrates these combined workflow issues:

> We found that [the labor and delivery area] was one of the most complex places in the
> hospital because the patients are going from the screening room to the pod room to a labor
> room to the delivery room to postpartum, and each of those [is] a different level of care
> and so orders need to be rewritten . . . and although nurses are good about blending the
> orders as necessary . . . the computer is trying to execute JCAHO rules about changing
> orders for every level of care [and they aren't] nearly so flexible.

The clinical ordering process might appear to follow predictable steps: For example, a clinician places an order (enters it through the computer), the system routes it to the desired destination, the order is processed, and the requested action occurs. In actual clinical practice, however, the process is much more adaptable, and includes a variety of checks, balances, interventions, and exceptions.[11] This noncomputerized process of placing an order is multithreaded, and consists of several concurrent and asynchronous steps, each of which may modify, terminate, or intervene in the processing of a given medical order. Many CPOE implementations change or eliminate these multiple interdependent steps, resulting in fewer process reviews and greater potential difficulties.

Our research team noted many instances where fewer process reviews led to problems. In one case, X-ray orders were unnecessarily duplicated:

> We probably underestimated the gatekeeper function that the clerical staff [provided] . . . One of the first symptoms is that patients had daily chest X-ray orders in many units, and the clerk had sort of provided a function of questioning after a certain amount of time, based on what he or she knew about the patient, [if] this was still appropriate. Once we automated those daily chest X-ray orders, [they] went on ad infinitum until we came up with an intervention to address that.

In another instance, a double-check was eliminated:

> The process before wasn't just the clerk writing down the allergies. The process was the clerk writing down the allergy and then the physician reviewing it before anyone took any action on it . . . that physician review was not kept in the [new CPOE] system.

One of the most often cited benefits of CPOE is the ability for clinicians to enter orders from anywhere in the hospital, or even from home. However, such new workflows can cause unexpected duplications or contradictions among orders, to the point of endangering patient care. For example, we heard this complaint:

> We had a lot more instances of within 30 seconds of each other, two—sometimes three—providers would enter the same order [from different locations,] and so it really forced us to go back and really do more education on being careful to look and see what's active before you enter a new order.

2.4.3 Type 3: Never-Ending Demands for System Changes

Never-ending system demands arise regarding hardware and software purchases, implementation tasks, and maintenance issues. They represent UACs from the perspective of both administration and IT staff. Implementation of CPOE requires advanced hardware platforms that can support clinical software. Purchase or upgrading of hardware is not a one-time event, as future technology advances make this an ongoing need. As clinical software systems increase in scope and capabilities, more users require more computer access time, via more computers. Clinicians complain when competition for access to computers interferes with accomplishing clinical tasks, particularly during the busiest hours (e.g., after morning rounds).

Software application demands are also never ending. One clinical development group moved upgrade releases from a weekly to monthly schedule because "the testing requirements . . . [were] becoming unbearable." In another instance, JCAHO recommendations to eliminate the use of common abbreviations meant that "there [were] over 4000 occurrences of the abbreviation "QD" in various order-entry templates that would have to be manually changed," and this was only one of a list of about 20 changes:

> I can't imagine how much work it is going to take to review all of the screens to find them, and what the incidence of new errors might be during the fix, such as eliminating an element in a pick list by accident, or making a typo in some drug name.

Overhead in maintaining systems and data increases regularly. When CPOE systems allow clinicians to create their own order sets, disparate single-user sets proliferate. In this scenario, it becomes progressively difficult to standardize, update, or maintain the order sets over time. It is also difficult to reconcile old order sets with new institutional initiatives to streamline care processes and to follow the most recent evidence-based clinical practice guidelines.

CPOE adoption transforms some acute problems into subtle and insidious ones. One author calls this phenomenon the "revenge effects" of technology.[12] During system implementation, the organization necessarily focuses on system go-live activities. But after the go-live event occurs, ongoing work begins. The system must be tuned, upgraded, tested, interfaced with other systems, and backed up regularly. As the clinical staff increasingly depends on the technology for their daily work, pressure to keep the system operational $24 \times 7 \times 365$ increases. Round-the-clock help-desk support becomes necessary. All employees must be trained in system use, and then retrained after substantive system changes occur. Backup systems must operate if the primary system fails. The burden on the technical support staff rarely levels off. Although these consequences can be anticipated, their extent is typically underestimated. One study participant referred to CPOE maintenance as "repairing a jet engine in flight" because the consequences of making mistakes with these systems are "orders of magnitude" greater than the consequences observed in less integrated, less closely coupled clinical systems.

As a CPOE system evolves, users rely more completely on the software and demand ever more sophisticated functionality for clinical support. As medical practices evolve, corresponding new features must be added to the original implementation. Over time, complex interactions among the numerous software features can make the installation both unmanageable and outdated, such that the system needs to be replaced with a newer (and "cleaner") version:

The fact that you develop a critical mass of code [means that] doing anything radically different becomes extremely difficult when you have an installed user base that you are supporting. So a lot of the early rapid flexibility and leeway you had in the early years of implementation you get stuck with . . . and it isn't easy to sort of wipe things clean and start over.

2.4.4 Type 4: Problems Related to Paper Persistence

Many CPOE vendors advertise products as helping an organization to "go paperless." While eliminating the paper-based medical record has clear advantages, including "improved legibility; simultaneous, remote access; and integration with other information sources,"[13] one should not confuse this concept with the premise of eliminating the use of paper by clinicians in their efforts to take care of patients. Instead, the key issue is to decrease or eliminate the dependency on ineffective, paper-based processes that form barriers to optimal healthcare delivery. In this regard, CPOE systems can be particularly effective.

Use of paper was endemic in the five institutions we visited as part of our study. This usage was especially pronounced when paper interfaces served as substitutes for electronic CPOE system integration with other clinical systems (e.g., medication administration recording, pharmacy dispensing, or laboratory ordering). In hospitals where CPOE and ancillary system integration was incomplete, we observed computerized orders being printed out in the processing departments, then reentered into the local department's CIS. We typically saw nurses manually transcribing allergy, blood type, and medication information from the CPOE system to paper-based medication administration records.

We observed providers using paper for temporary, handwritten data storage for later entry into the computer and, conversely, as a portable, disposable, computer output display modality for quick reference use during their workdays. In hospitals where CPOE systems generated and printed patient summary sheets, we noticed some providers documenting patient progress notes on these printouts. Despite an explicit directive on CPOE printouts not to do so, and counter to current recommendations by the American Hospital Information Management Association,[14] clinicians placed annotated printouts in the patient's chart as formal documentation. These are but a few examples of paper persistence—and often proliferation—in all areas of patient care.

By contrast, paper often serves as a necessary, sometimes superior, cognitive memory aid. As one clinician noted, "I like to have the information on paper where I can hold on to it." Furthermore, paper remains the most malleable, flexible, and easily transportable data medium available. Organizations are understandably hard pressed to limit its use. Personnel need simply point and click to print "hard copies" of stored information. Although paper-based clinical record storage will become obsolete, use of paper in the clinical setting will not. One leader indicated that his institution uses roughly "1.6 million pieces of paper per month—printed or copied —and we think half is related to clinical care . . . we print and destroy 40% of that paper."

2.4.5 Type 5: Untoward Changes in Communication Patterns and Practices

CPOE systems often dramatically alter traditional communication patterns among care providers, ancillary services, and clinical departments. Installation of CPOE replaces the nexus of previously interpersonal conversations regarding provision of care with a computer system. Some describe CPOE as providing an "illusion of communication"[15] because it promotes the belief that entry of an order into the system ensures that the proper people will see it and act upon it. This unfounded belief is especially problematic for "stat" (emergent) orders because their execution in a timely fashion should depend on interpersonal communication (even post CPOE implementation). Many CPOE users assume that electronic transmission will be efficient, however, and do not understand that fast computer or network transmission does not guarantee fast, or accurate, notification of the person who must take care of the order. In our study, we observed many instances where emergent orders were not only placed in the CPOE system, but also (redundantly) phoned in to assure they took place immediately.

Doctors, nurses and other providers consistently report that clinical systems such as CPOE can cause unsatisfactory reductions in face-to-face communication regarding patient care. The providers further suggest that reduction in communication increases the likelihood of errors due to miscommunication, delayed initiation and execution of orders, and fewer team-wide discussions regarding planning and coordination of care. For example, order-entry sessions may precede or remotely follow ICU rounding sessions, when the attending and consulting physicians, the respiratory therapist, and the nurse come together to discuss patient care. Rapid and significant changes in the patient's condition in the time interval between order discussion and order entry may lead to omission or delayed entry of some relevant orders. CPOE systems can exacerbate problems related to use of verbal orders in conjunction with system entry; some institutions have gone to the extreme of banning verbal orders except in the case of emergencies:

> It is not uncommon for a physician to enter an order which has also been verbally stated to the nurse. The nurse, acting on that verbal order, then goes back to, say, mix an IV. The doctor in the meantime changes his mind [about the IV admixture, then places the order] and does not tell the nurse until the bag has been hung, resulting in a waste of a $100 bag of IV fluid.

2.4.6 Type 6: Negative Emotions

Organizational change is never easy, and shifting clinical practices and workflows can engender enormous emotional resentment in end users. Sittig et al. have suggested that "a specific event or series of events that either cause the person to succeed or fail in reaching his or her goal(s)" triggers many emotions and that such emotions can affect one's ability to carry out complex physical and cognitive tasks.[16] Shifting from paper-based order generation to CPOE is bound to evoke strong emotional responses as users struggle to adapt to the new technology.

In our study, we noted a wide variety of emotional responses to CPOE, including both strongly negative and highly positive emotions. Negative comments predominated. The amount of time a CPOE system had been in use strongly correlated with the level of positive emotions the system elicited. For example, one nurse described her first impression of a CPOE system in this manner:

> At first we hated every second of it. I mean we were all like, "I have sick patients here. I'm busy. I don't have time to sit here [at the computer] for 20 minutes." It was a pain.

Most agree that the high level of negative emotions decreases over time: "It gets better."

2.4.7 Type 7: Generation of New Kinds of Errors

Study results indicated that CPOE adoption can generate new kinds of healthcare practice-related errors. Previous studies described roles for CPOE in both preventing and causing medication errors.[4,17–22] Here, we focus on new types of errors that emerge when CPOE replaces paper-based ordering.

New CPOE-related errors result from: problematic electronic data presentations; confusing order-option presentations and selection methods; inappropriate text entries; misunderstandings related to test, train, and production versions of the system; and workflow process mismatches. System designs (including those characterized by poor data organization and data omissions) and end-user confusion about system functionality also contribute to new forms of errors. When users make data-entry selections from pick lists (drop-down lists), a new class of "juxtaposition errors" results from making a wrong selection without realizing it. For example, long, dense pick lists predispose a provider to selecting a patient name adjacent to the intended name. Ideally, the CPOE system should provide adequate feedback on who was selected (e.g., displaying the selected name in large letters on the next screen). If this does not occur, the user may proceed to enter an entire set of orders for the wrong patient. "Backing out" of such erroneous orders before they are executed can be problematic. Similar errors occur whenever pick lists facilitate selection of orderable item parameters.

CPOE systems manage massive amounts of clinical information. However, CPOE workstation screens cannot display large amounts of data simultaneously. Thus clinicians must learn to navigate serially through CPOE interface screens to perform their work. When busy clinicians cannot readily find the "correct" data entry location, they tend to enter data where it *might* fit, such as in a "miscellaneous" section. Although such information resides in the system, it may be stored in a manner that makes categorizing, cross-checking, processing, and acting upon it more difficult. Furthermore, improper data placement may impede other clinicians from finding important information:

> The biggest problem with orders [is that] people get frustrated finding the right spot to put something, or don't see what they need immediately, then end up entering orders in the miscellaneous section. This makes it easy to miss things and hard to capture data on the orders being entered.

Poor coordination in deploying test, training, and production versions of CPOE systems can create new kinds of errors. For example, unless safeguards are in places that allow only obviously artificial patient names in "test" and "train" modes, it may be possible for a clinician to use a test system for a patient name that by chance matches an actual patient. The user would not know that the orders entered *will not* be acted upon, because they were placed outside the production system. Similarly, without appropriate safeguards, a provider might, during a training session, enter orders for a "test" patient who is actually a "live" patient in the production system. The resulting "test" orders *will* be processed and have consequences. Finally, problems may emerge when test patients are identified by only a cute, simple name (e.g., Tom E. Test, rather than a safer ZZZTest, ZZZTest), especially when an actual patient with a last name of "Test" is admitted.

2.4.8 Type 8: Unexpected and Unintended Changes in Institutional Power Structure

As CPOE systems enforce specific clinical practice patterns, while at the same time monitoring clinicians' behaviors, they induce changes in the power structure and culture of an organization. Power, whether formally or informally delegated, plays an extremely important role in any work environment.[23] This is especially true in health care, where lines of authority emanate from a tradition based on educational hierarchies, various providers' roles, differences between general practitioners and specialists, and more.[24] For these reasons, adoption of CPOE systems may subtly bring important power issues to the surface.

CPOE system configurations control who may do what (and when) through the use of clinical, role-based authorizations. While narrowly defined authorizations may lead to much needed role standardizations that reduce unnecessary overlap in clinical practice, the same constraints may also redistribute work in unexpected ways, causing frustration. Physicians may resent the need to enter orders directly into a computer. They may view this procedure as a traditionally clerical task. Nurses may refuse to take verbal orders except in cases of emergency or insist that the physician enter orders into the CPOE system before any order will be carried out. We heard many nurses express sentiments such as, "It really isn't a nursing job to convince the doctor to use the system—and if we must use it, they should, too."

Physicians report loss of professional autonomy when CPOE systems prevent them from ordering the types of tests or medications they prefer, or force them to comply with clinical guidelines they may not embrace, or limit their narrative flexibility through structured rather than free-text clinical documentation. It is irritating when the system directly rewords clinical orders to standardize them:

> I didn't realize how important nomenclature was in ordering . . . Our system has a very robust synonym process where you can use synonyms; we have synonyms for . . . common misspellings. The doctor puts [an order] in and [it] goes into the patient's chart with the kind of "accepted" name. They'll come back later and look at that and say, "I didn't order that. It doesn't look like anything I ordered."

In the preceding example, the clinician might object to the forced use of awkward terminology. Such terminology not only disrupts clinicians' workflows, but might also compromise patient care. While order standardization may benefit the organization, it may confuse clinicians. This represents a nontrivial domain in which clinicians' professional autonomy may be circumvented.

Whether the power base is centralized or decentralized plays an important role in occurrence of UACs. Centralized power structures use top-down, hierarchical formats to mandate compliance with organizational rules and to enforce procedure standardization. Decentralized arrangements lead to greater variations in CPOE system configuration and utilization, and increase competition and conflict among departments. Conflicts create significant problems for IT departments and lead to problems with the application consistency, clinical coordination, and evaluation of impact on patient care.

When many departments participate in CPOE implementation, significant unanticipated power shifts occur. Viewed as the new enforcer of standards, the IT department gains power, even when other departments mandate the standards. This can be frustrating to others:

> CPOE should be a clinical project, not an IT project, but it's still amazing how much I think it comes out of the IT department.

Administration and quality assurance departments, in turn, gain power by requiring users to comply with CPOE-based directives:

> The doctor can't place that order unless he fills in that field. They're [administration] really happy, and I think that's really a problem for the doctors.

2.4.9 Type 9: Overdependence on Technology

As CPOE technology diffuses and becomes entrenched within organizations, clinical care delivery becomes inextricably dependent upon it.[25,26] System failures increasingly wreak havoc when paper backup systems are not in place:

> It's funny now. When the system goes down, we don't remember how to work with paper.

Prolonged system failures (lasting hours) can so dramatically halt the flow of clinical information that outpatient activities may be curtailed or canceled, and emergency rooms at trauma centers may divert admissions until vital systems are restored. The more widely and deeply diffused the technology, the more difficult it becomes to work without it.

Although limited experience exists, embedding clinical decision support (CDS) within CPOE systems may increase clinician-users' access to educational material[27,28] and may affect learning and retention.[29] Clinicians who have only worked with CPOE systems using CDS technologies face new and interesting problems when they transfer to work settings *without* this technology. Important knowledge gaps might emerge for the clinician who relied on CDS to provide real-time information and/or error prevention. Such a clinician might have trouble remembering standard dosages, hospital formulary recommendations, and medication contraindications. Conversely, prior use of CDS may actually promote learning, through repeated and consistent presentation of the sorts of information mentioned in real-time alerts and reminders. Thus, depending on an individual's learning style and the type and amount of decision support available, CDS may actually enrich clinical training.

Finally, consider this comment: "Our society is geared so that if it's in the computer, it must be accurate and complete, and as we know, it just isn't so." For example, certain free-text fields cannot be processed by CDS components of CPOE. Allergies to medications, when entered as atypical abbreviations ("pcn," "smtx"), are a common example. Clinicians assume the computer "knows" the information—which can be an especially worrisome concern with clinically vital information such as allergies.

2.4.10 Decision Support Systems

During analysis of unintended consequences, researchers learned that clinical decision support generates a disproportionately large number of UACs (more than 25%), spanning all nine types of UACs profiled in this chapter. Clinical decision support functionality, which is viewed as necessary to make CPOE beneficial, often lacks relevance to many specific clinical situations. Accordingly, CPOE-based CDS is not consistently useful for clinicians, is impractical to maintain, and is less than fully reliable in complex situations.[16,21,30–35]

2.5 Discussion

This study identified types of negative (adverse) unintended consequences resulting from CPOE implementation. Project members observed end users interacting with CPOE, interviewed key players involved in implementations, and analyzed transcripts of meetings designed to elicit UAC-related information. The project identified hundreds of UACs and grouped them into nine categories. Project members came to the realization that the degree of undesirability of each UAC depends to a great extent on one's perspective. We found that the nine types of unintended consequences emerged with surprising regularity at the sites visited as part of the study, and many UACs originated with attempts to implement clinical decision support.

The project resulted in several significant lessons. First, UACs occur during all CPOE implementations, though not all institutions experience all of the types of UACs. Second, the nine types of UACs occur in a widespread manner. These UACs pose significant consequences for clinicians, technical staff, and organizations as a whole. Finally, the project's typology establishes a framework for systematic approaches to addressing of these issues.

2.5.1 More Work/New Work

CPOE systems can significantly increase clinician workload, and improved system design may not reduce the amount of new work such systems require. Great care must be taken to balance the risk of over-alerting against the risk of not alerting. Developers should rework clinical system interfaces to (1) reduce collection of redundant information, (2) display relevant information in logical locations, and (3) reduce the amount of required typing. The lesson is that more work for the clinician is inevitable, and constitutes an issue that must be addressed in the planning process. Successful implementations balance required new work with system-based reductions in old work to make use of the systems by clinicians tolerable.

2.5.2 Workflow

Clinical workflows are complex, and clinical computer technology integration significantly affects healthcare workflows.[9,35–49] Modeling clinical workflows is difficult because clinical practice is so inherently complex, interruption driven, and constantly changing. Thus no CPOE system will ever perfectly fit all workflows of a given hospital, all of the time. Even if a system initially did so, a lesson learned is that it would not eliminate the need for constant future adaptation to changing workflows. Whenever there are adjustments, there will be unintended consequences.

2.5.3 Never-Ending System Demands

CPOE systems evolve (e.g., are reconfigured, enhanced, or replaced) over time, making hardware and software upgrades necessary. With each change, implementers should expect unintended consequences. As changes occur, users must be retrained and quality assurance measures must be reassessed. The lesson is that planning must allocate adequate resources for ongoing improvements.

2.5.4 Paper Persistence

While electronic medical record systems trend toward "going paperless," healthcare organizations, as a whole, do not. Vendors and administrators alike must understand differences between having a paperless record system and a paperless office. Organizations must delineate what constitutes legal documentation in the presence of an electronic medical record.[50] A likely reduction in paper use will follow full integration of disparate clinical information systems and widespread deployment of clinical computing workstations or wireless, hand-held devices. Paper is here to stay for utilitarian—as opposed to permanent record-keeping—purposes. Attempts to limit its pragmatic use in health care are often misguided.

2.5.5 Changes in Communication Patterns and Practices

Computerized systems are unlikely to replicate the richness of face-to-face communication, but computer-based communication systems must improve. Improvements in system interface design must pay special attention to the communication needs of healthcare providers. The lesson is that CPOE implementation changes clinical communication patterns. In addition, a comprehensive communication plan that reaches all levels of the organization must be part of any CPOE project management plan.

2.5.6 Emotions

Emotional responses to change are inevitable. These responses can point out significant problems with the system design, and ultimately can lead to solutions of those problems. Training and open communication can help to promote better understanding, which may reduce the negative emotional responses to CPOE.

2.5.7 New Kinds of Errors

CPOE systems prevent some types of errors while creating or propagating new ones.[4,5,18,19,21,22,30,51–58] Many new errors result from straightforward system interface design problems, such as dense pick lists that cause juxtaposition problems. Recognizing current unintended consequences should encourage system designers to optimize the human–computer interface design, and to exert caution when implementing new alerts.

2.5.8 Changes in the Power Structure

Because CPOE-related power changes affect organizational and personal autonomy, they often cause significant UACs for end users. Most often it is the physician who loses power—a point that must be recognized and dealt with explicitly during the CPOE planning process.

2.5.9 Overdependence on Technology

Health care is increasingly dependent on technology, and this situation is unlikely to change. Dependence on technology must never become so great that basic medical care cannot be provided in its absence. Planning for management of unexpected downtime is critical.

2.5.10 Decision Support Systems and CPOE

The project described in this book identified decision-support-based clinical alerts as a common source of UACs. Inevitably, there are both optimal and less good approaches to designing and implementing decision support. The suboptimal approaches to CDS can have widespread negative effects on clinical practices. Poorly designed alerts constantly interrupt providers, often with trivial, redundant, or already known information. Alerts present a major workflow process issue, adding to the steps required to enter an order. In addition, clinicians report inappropriate alerts to be highly frustrating nuisances.

Alerts can change the power structure both beneficially and adversely, and they present clear challenges to professional expertise or autonomy. Because alerts appear to be from "the system," they may be viewed as correct when they are erroneous. Conversely, when appropriate alerts are ignored, error prevention may not occur, yet redundant or unnecessary test avoidance diminishes. Building and maintaining an appropriate set of CPOE alerts based on an up-to-date evidence base is an onerous, never-ending task.

2.5.11 Looking to the Future

Improvements are both warranted and attainable with respect to unintended consequences. Currently, CPOE technology is immature and rapidly evolving. The field
must prioritize its ongoing efforts to better understand the nature of clinical work.
Designers must build systems that respect healthcare providers' burdens, and constant feedback from providers should guide system implementation and evolution.
Workflow improvements require experience and time. As the adoption of CPOE
systems increases, so will the wealth of knowledge about how to use the systems to
improve care. Improved interface designs may eliminate or reduce the possibility of
juxtapositions and other related errors. As technology evolves, dramatic improvements
in providing clinicians with pertinent data will emerge. Such technological advances
first and foremost require clinical involvement to better support clinical work. Only
through a careful combination of CPOE-related research, design, feedback, and understanding will many of the unintended consequences be reduced or eliminated.

2.6 Conclusion

It is important to view UACs as the result of a constellation of factors. Their causative
agents almost never occur in isolation. Likewise, UACs tend to emerge in all aspects
of health care, whether or not the task from which they emerge is technological.
Not all types of UACs may occur with every CPOE implementation. The goal is to
discover and understand causative factors leading to UACs, thereby allowing future
CPOE developers and implementers to mediate or eliminate UACs that are preventable or remediable. Doing so will reduce the negative impacts CPOE systems have on
providers, patients, and administrators. It is only through careful evaluation of UACs
that we can gain better insight into how to best approach these problems.

2.7 References

1. Committee on Quality Health Care in America. *To err is human: Building a safer
   health system.* Washington, DC: Institute of Medicine; 1999.
2. Committee on Quality Health Care in America. *Using information technology:
   Crossing the quality chasm: A new health system for the 21st century.* Washington,
   DC: Institute of Medicine; 2001.
3. Leapfrog Group. Factsheet: Computer physician order entry. Available at: http://
   www.leapfroggroup.org/for_hospitals/leapfrog_safety_practices/cpoe. Accessed
   August 12, 2009.
4. Koppel R, Metlay JP, Cohen A, Abaluck B, Localio AR, Kimmel SE, et al. Role
   of computerized physician order entry systems in facilitating medication errors.
   JAMA. 2005;293(10):1197–1203.
5. Ash JS, Berg M, Coiera E. Some unintended consequences of information
   technology in health care: the nature of patient care information system-related
   errors. *J Am Med Inform Assoc.* 2004;11(2):104–112.
6. Rogers EM. *Diffusion of Innovations.* 5th ed. New York: Free Press; 1998.

7. Ash J, Sittig D, Dykstra R, Guappone K, Carpenter J, Seshadri V. Categorizing the unintended sociotechnical consequences of computerized physician order entry. *Int J Med Informatics.* 2007 Jun;76 Suppl 1:S21–7.

8. Poissant L, Pereira J, Tamblyn R, Kawasumi Y. The impact of electronic health records on time efficiency of physicians and nurses: a systematic review. *J Am Med Inform Assoc.* 2005;12(5):505–516.

9. Overhage JM, Perkins S, Tierney WM, McDonald CJ. Controlled trial of direct physician order entry: effects on physicians' time utilization in ambulatory primary care internal medicine practices. *J Am Med Inform Assoc.* 2001;8(4):361–371.

10. Scott JT, Rundall TG, Vogt TM, Hsu J. Kaiser Permanente's experience of implementing an electronic medical record: a qualitative study. *Br Med J.* 2005;331(7528):1313–1316.

11. Hazlehurst B, McMullen C, Gorman PN, Sittig DF. How the ICU follows orders: care delivery as a complex activity system. *Proc AMIA Annu Symposium.* 2003;284–288.

12. Tenner E. *Why Things Bite back: Technology and the Revenge of Unintended Consequences.* New York: Random House; 1996.

13. Powsner SM, Wyatt JC, Wright P. Opportunities for and challenges of computerisation. *Lancet.* 1998;352(9140):1617–1622.

14. AHIMA e-HIM Work Group on Health Information in a Hybrid Environment. *The complete medical record in a hybrid EHR environment. Part III: Authorship of and printing the health record* (AHIMA practice brief). Available at: http://library.ahima.org/xpedio/groups/public/documents/ahima/bok1_021583. hcsp?dDocName=bok1_021583. Accessed August 12, 2009.

15. Dykstra R. Computerized physician order entry and communication: reciprocal impacts. *Proc AMIA Annu Symposium.* 2002;230–234.

16. Sittig DF, Krall M, Kaalaas-Sittig J, Ash JS. Emotional aspects of computer-based provider order entry: a qualitative study. *J Am Med Inform Assoc.* 2005;12(5): 561–567. (Available as Chapter 3 in this book.)

17. Bates DW, Kuperman GJ, Rittenberg E, et al. A randomized trial of a computer-based intervention to reduce utilization of redundant laboratory tests. *Am J Med.* 1999;106(2):144–150.

18. Bates DW, Teich JM, Lee J, et al. The impact of computerized physician order entry on medication error prevention. *J Am Med Inform Assoc.* 1999;6(4):313–321.

19. Bobb A, Gleason K, Husch M, Feinglass J, Yarnold PR, Noskin GA. The epidemiology of prescribing errors: the potential impact of computerized prescriber order entry. *Arch Intern Med.* 2004;164(7):785–792.

20. Horsky J, Kuperman GJ, Patel VL. Comprehensive analysis of a medication dosing error related to CPOE. *J Am Med Inform Assoc.* 2005;12(4):377–382.

21. Kaushal R, Shojania KG, Bates DW. Effects of computerized physician order entry and clinical decision support systems on medication safety: a systematic review. *Arch Intern Med.* 2003;163(12):1409–1416.

22. Oren E, Shaffer ER, Guglielmo BJ. Impact of emerging technologies on medication errors and adverse drug events. *Am J Health-System Pharm.* 2003;60(14):1447–1458.

23. Robbins SP. *Organizational Behavior*. 11th ed. Upper Saddle River, NJ: Pearson Prentice Hall; 2005.

24. Massaro TA. Introducing physician order entry at a major academic medical center: 1. Impact on organizational culture and behavior. *Academic Med.* 1993;68(1):20–25.

25. Kilbridge P. Computer crash: lessons from a system failure. *N Engl J Med.* 2003;348(10):881–882.

26. Oppenheim MI, Vidal C, Velasco FT, et al. Impact of a computerized alert during physician order entry on medication dosing in patients with renal impairment. *Proc AMIA Annu Symposium.* 2002;577–581.

27. Rosenbloom S, Geissbuhler A, Dupont W, et al. Effect of CPOE user interface design on user-initiated access to educational and patient information during clinical care. *J Am Med Inform Assoc.* 2005;12(4):458–473.

28. Bochicchio GV, Smit PA, Moore R, et al. Pilot study of a web-based antibiotic decision management guide. *J Am Coll Surg.* 2006;202(3):459–467.

29. Knight A, Kravet S, Harper G, Leff B. The effect of computerized provider order entry on medical student clerkship experiences. *J Am Med Inform Assoc.* 2005;12(5):554–560.

30. Galanter WL, Polikaitis A, DiDomenico RJ. A trial of automated safety alerts for inpatient digoxin use with computerized physician order entry. *J Am Med Inform Assoc.* 2004;11(4):270–277.

31. Handler JA, Feied CF, Coonan K, et al. Computerized physician order entry and online decision support. *Academic Emerg Med.* 2004;11(11):1135–1141.

32. Hsieh TC, Kuperman GJ, Jaggi T, et al. Characteristics and consequences of drug allergy alert overrides in a computerized physician order entry system. *J Am Med Inform Assoc.* 2004;11(6):482–491.

33. Kawamoto K, Lobach DF. Clinical decision support provided within physician order entry systems: a systematic review of features effective for changing clinician behavior. *Proc AMIA Annu Symposium.* 2003;361–365.

34. Neilson EG, Johnson KB, Rosenbloom ST, et al. The impact of peer management on test-ordering behavior. *Ann Intern Med.* 2004;141(3):196–204.

35. Sengstack PP, Gugerty B. CPOE systems: success factors and implementation issues. *J Healthcare Inform Manage.* 2004;18(1):36–45.

36. Berg M. Medical work and the computer-based patient record: a sociological perspective. *Meth Inform Med.* 1998;37(3):294–301.

37. Berg M. Patient care information systems and health care work: a sociotechnical approach. *Int J Med Informatics.* 1999;55(2):87–101.

38. Beuscart-Zephir MC, Pelayo S, Degoulet P, Anceaux F, Guerlinger S, Meaux JJ. A usability study of CPOE's medication administration functions: impact on physician–nurse cooperation. *Medinfo.* 2004;11(Pt 2):1018–1022.

39. Briggs B. The top 10 CPOE challenges. *Health Data Manage.* 2004;12(7):20–22.

40. Cheng CH, Goldstein MK, Geller E, Levitt RE. The effects of CPOE on ICU workflow: an observational study. *Proc AMIA Annu Symposium.* 2003;150–154.

41. Cordero L, Kuehn L, Kumar RR, Mekhjian HS. Impact of computerized physician order entry on clinical practice in a newborn intensive care unit. *J Perinatol.* 2004;24(2):88–93.

42. Dourish P, Bellotti V. Awareness and coordination in shared work spaces. In: *ACM Conference on Computer-Supported Cooperative Work (CWSW '92)*. Toronto, Canada: ACM Press; 1992:107–114.

43. Eisenberg F, Barbell AS. Computerized physician order entry: eight steps to optimize physician workflow. *J Healthcare Inform Manage*. 2002;16(1):16–18.

44. Gaillour FR. Why do health systems flop with CPOE? Ask Yoda. *Physician Exec*. 2004;30(2):28–29.

45. Levick D, Lukens HF, Stillman PL. You've led the horse to water, now how do you get him to drink: managing change and increasing utilization of computerized provider order entry. *J Healthcare Inform Manage*. 2005;19(1):70–75.

46. Poon EG, Blumenthal D, Jaggi T, Honour MM, Bates DW, Kaushal R. Overcoming barriers to adopting and implementing computerized physician order-entry systems in U.S. hospitals. *Health Affairs*. 2004;23(4):184–190.

47. Ali NA, Mekhjian HS, Kuehn PL, et al. Specificity of computerized physician order entry has a significant effect on the efficiency of workflow for critically ill patients. *Crit Care Med*. 2005;33(1):110–114.

48. Shane R. CPOE: the science and the art. *Am J Health-System Pharm*. 2003;60(12):1273–1276.

49. Shu K, Boyle D, Spurr C, et al. Comparison of time spent writing orders on paper with computerized physician order entry. *Medinfo*. 2001;10(Pt 2):1207–1211.

50. AHIMA e-HIM Work Group on the Legal Health Record. Update: guidelines for defining the legal health record for disclosure purposes. *J AHIMA*. 2005;76(8): 64A–64G.

51. Berger RG, Kichak JP. Computerized physician order entry: helpful or harmful? *J Am Med Inform Assoc*. 2004;11(2):100–103.

52. Hagland M. Safe ways: hospitals looking to improve patient safety are turning to CPOE, bar coding, and e-prescribing. *Healthcare Informatics*. 2004;21(8):20–25.

53. King WJ, Paice N, Rangrej J, Forestell GJ, Swartz R. The effect of computerized physician order entry on medication errors and adverse drug events in pediatric inpatients. *Pediatrics*. 2003;112(3 Pt 1):506–509.

54. Kremsdorf R. CPOE: not the first step toward patient safety. *Health Manage Technol*. 2005;26(1):66.

55. Nebeker JR, Hoffman JM, Weir CR, Bennett CL, Hurdle JF. High rates of adverse drug events in a highly computerized hospital. *Arch Intern Med*. 2005;165(10): 1111–1116.

56. Potts AL, Barr FE, Gregory DF, Wright L, Patel NR. Computerized physician order entry and medication errors in a pediatric critical care unit. *Pediatrics*. 2004;113(1 Pt 1):59–63.

57. Scanlon M. Computer physician order entry and the real world: we're only humans. *Joint Commission J Quality Safety*. 2004;30(6):342–346.

58. Wears RL, Berg M. Computer technology and clinical work: still waiting for Godot. *JAMA*. 2005;293(10):1261–1263.

3

A Sociotechnical Assessment of Unintended Adverse Consequences Associated with Clinical Workflow

Based on: Campbell EM, Guappone KP, Sittig DF, Dykstra RH, Ash JS. Computerized provider order entry adoption: implications for clinical workflow. *J Gen Intern Med.* 2009;24(1):21–26.

3.1 Key Points

- CPOE systems can affect clinical work by (1) introducing or exposing ergonomic and human–computer interaction problems; (2) altering the pace, sequencing, and dynamics of clinical activities; (3) providing only partial support for the work activities of all types of clinical personnel; (4) reducing clinical situation awareness; and (5) poorly implementing organizational policies and procedures.
- The five kinds of unintended adverse consequences related to workflow with CPOE can be mitigated by iteratively altering both clinical workflow and the CPOE system until a satisfactory fit is achieved.

3.2 Introduction

3.2.1 Defining Clinical Workflow

Workflow can be defined as the processes or procedures needed to produce or modify work, products, or the delivery of services. Processes usually have discrete beginning and ending points and may consist of multiple tasks, which, when combined, represent logical units of work. These processes are guided by a set of conditions that specify input and output requirements. Input requirements outline what is needed for the process to begin; output requirements describe what the result of a completed process might be. Additional process specifications might prescribe rules for the sequencing of subtasks, including what needs to be done in what order and whether some tasks can be done in parallel or the level of training needed for performing a task and who has permission to do it. Finally, processes identify the resources needed for task completion, including people, information, money, or other necessary tools. Workflow processes may involve varying levels of automation or computerization, depending on the desired outcome. By definition, then, clinical workflow is the set of processes or procedures, people, and tools, such as CPOE, that are related to the delivery of patient care.

3.2.2 A Sociotechnical Conceptualization of Clinical Workflow

The sociotechnical approach considers clinical workflow practices as *systems* in which the social and technical aspects of the work are fundamentally interrelated such that they are not separable into discrete domains for study. Berg has called the social and organizational aspects of work "[the] politically textured process of organizational change, in which users have to be put at center-stage."[1] This "socio" part of the socio-technical approach is inextricably bound to the more formal, structured, and quantifiable technical features of information system design, adoption, and evaluation.[1-3]

Sociotechnical approaches consider clinical work practices to be heterogeneous networks of social and technical resources with interrelated functions that focus primarily on the delivery of health care.[1,3-5] Thus, under this framework, a technology solution (such as a clinical application) represents one or more nodes in a network of care delivery, operating within environmental constraints affecting both its functionality and the environment in which it is embedded. Using a sociotechnical approach, we can better appreciate how systems of work practice react when technology is added by evaluating the new network behaviors that emerge as a result of the change. This analysis, in turn, helps us study the fit between the technology and the network of care so that we can find ways to improve both iteratively.

3.2.3 Evaluating Sociotechnical Systems

Traditional empirical evaluation methods can be ineffective for improving our understanding of these complex sociotechnical systems. Kaplan, in an excellent overview of clinical systems evaluation methods, suggests that "concerns with such contextual questions as power, culture, group relationships, work routines, stakeholders, professional values, social networks, institutional organization, and judgment elude quantitative and [randomized control]-type evaluation approaches."[6] For example, a time–motion study may demonstrate a quantifiable change in the time it takes clinicians to perform certain work activities before and after a CPOE implementation. However, such a study will not provide any insight into the reasons for user dissatisfaction with certain system elements. Low user satisfaction might actually be the root cause of the increased time required to complete the task; it might also explain the variability in the amount of time it takes different users to complete the same task. The root cause for the low satisfaction might be that clinicians feel that their personal autonomy is being threatened.[1,4] The time–motion study is valuable in its own right, but the only way to understand the layers of complexity and interrelatedness of the sociotechnical system and to discover why the change in time occurs is to conduct a qualitative sociotechnical evaluation.[1,4,7-10] This approach offers insight into the "fit"[6] between such system elements as the user, the environment, and the technology, and it illuminates areas for improvement when unintended adverse consequences of CPOE implementation emerge.

Furthermore, when using the sociotechnical approach, one should not (or perhaps cannot) separate the technology, as originally programmed by its developers, from its implementation. Even exquisitely designed and coded software can be implemented poorly. From the standpoint of the interaction between the software and clinical work-flow, it is only the use of the software by clinicians that is important and observable; an evaluation of a CPOE system must directly consider its implementation.

3.3 Key Findings

In our study, we identified five main types of unintended adverse consequences related to clinical workflow.

3.3.1 CPOE Raises Many Ergonomic and Human–Computer Interaction Issues

During our study, we identified many situations in which work activities were disrupted simply by the introduction of the computer into the work environment. These disruptions were particularly noticeable during the order-entry process. Such problems may arise from simple human–computer interface issues (e.g., poor user keyboarding skills, lack of computer experience or training, or poor user interface design). Some disruptions arise when work environments designed prior to the computer era cannot adequately accommodate the necessary hardware or provide sufficient workspace for the people who use it.

We also observed repeated instances of physical space limitations and ergonomic mismatches that interfered with users' work. For example, computer workspaces designed for use while standing up were excellent for retrieving information but proved difficult to use when simultaneously accessing paper documentation (such as a patient chart) or performing other tasks (such as using a mouse to highlight and select an item from a cascading menu). In such cases, clinicians need a place to sit, support for their arm or wrist, and a place to lay the paper chart while using the computer. One physician summed up this issue quite succinctly: "A computer that doesn't have a place to put the chart down is no computer I am willing to use." In addition, clinical work areas are notoriously busy and crowded, and competition for computers can be quite high, especially after morning rounds. Often, space constraints can create awkwardness, as described in our field notes:

> The team is using one work area about the size of a phone booth. The intern is at the PC and really is the only one entirely in the "booth." The resident and med student are crammed behind him, looking over his shoulders at the screen. The intern is reviewing a list of meds with the others, using the mouse as a pointer.

We noticed many issues related to poor CPOE system design, including overly cluttered screens or poor use of available display space, inconsistent application functionality, poorly sorted lists (or lists that would not sort at all), difficulty in finding where to chart or locate necessary clinical information, lack of convenient system defaults, and lack of appropriate safeguards to prevent selecting the wrong patient or entering incorrect data, to name just a few. One of our team members observed this simple example of poor design:

> I notice that the resident has to perform four mouse clicks to access an element on a list: (1) click on the pick list, (2) open the list with the down arrow, (3) select an item from the list, and (4) hit the return key to exit the pick list. Normally, this wouldn't be much of a problem, but the list contains only one element!

End users struggle with many human–computer interface and ergonomic issues when moving from a pen-and-paper information system that is flexible and highly portable to one that is much more rigid and fixed. Poor system interface design (e.g., overly complex screens, inconsistencies in the interface, poor grouping of like terms) can exacerbate this transition. In addition, placing computers in environments not originally designed for them often compounds user frustration due to limited accessibility (e.g., when there are insufficient numbers of computers for the staff) or lack of space (e.g., when computers consume available counter space). From a sociotechnical perspective, these issues can cause users to invent system workarounds, avoid thorough documentation, find other staff to serve as proxies to enter orders, or refuse to use the system altogether.

Taken as a whole, these issues highlight the importance of determining appropriate computer-to-staff ratios to assure sufficient hardware is available, along with paying careful attention to usability in interface design, to help increase both user acceptance and effective use of the system. These issues are not unique to CPOE systems, of course. Also, although they are well-recognized, there is room for significant improvement in current systems. We recommend more quantitative and qualitative usability studies to address where ergonomic issues and poor human–computer interface design directly affects clinical workflow, regardless of the clinical information system being evaluated. These studies would ideally be performed on an ongoing basis to assure that as systems are upgraded and as workflow changes for non-system-related reasons, the system continues to fit the workflow well. In addition, we look forward to the development of explicit interface design and usability criteria that must be met to certify CPOE systems through such organizations as the Certification Commission for Healthcare Information Technology (http://www.cchit.org).

3.3.2 CPOE Changes Work Pace, Sequence, and Dynamics

Implementing CPOE systems in the clinical setting disrupts workflow processes by altering the pace, sequence, and dynamics of clinical work practices. CPOE creates new work that fundamentally alters the work pace. For example, with nonmobile CPOE systems, providers must locate a computer, log into the system, select a patient, locate the appropriate screens to enter or view information, and log off once their work is complete. While logged into the system, providers may find it particularly difficult to access patient information housed in disparate clinical systems, especially when these other systems are not integrated with CPOE, require separate system logins, or cannot be accessed simultaneously. A physician explained:

> This is one thing that is annoying. For me to get lab values, I would have to exit out of the discharge summary, [look up the lab values,] then bring [the discharge summary] up again. It is just easier for me to look up values on a separate computer.

CPOE can further affect work pacing by requiring the provider to adopt new techniques for ordering simply to match data-entry requirements. For example, providers are usually prevented from writing with free text or using time-saving shorthand (such as abbreviations or acronyms); instead, they must learn to select items by navigating menus or selecting from long pick lists. In addition, clinical decision support in the form of clinical alerts can become a time-consuming annoyance when the alerts contain patient information that is already known, is incorrect, or is irrelevant.

CPOE systems can also interfere with the sequencing of clinical activities. Because these systems often model clinical work activities in linear patterns, important redundant steps to assure quality care can be eliminated. Likewise, the overly rigid modeling of workflow can eliminate much of the iterative and conversational nature of clinical activity important for the review and refinement of treatment plans. As a consequence, rigid implementation of rules, despite carefully planned interventions, can result in unexpected process failures. One clinician noticed, "One problem had to do with [inadvertently and automatically] discontinuing antimicrobials when the duration they were ordered for had elapsed."

Work can also be disrupted when rigid CPOE systems operate in clinical environments that must allow for some "fuzzy" care boundaries, as during transitions in level or location of care. For example, it is quite common for an admitting clinician to begin to write orders for an emergency department patient prior to transfer to an inpatient bed. Because some CPOE implementations associate orders with a patient's physical location, however, the system may prevent the admitting clinician from entering these orders, making it especially difficult for discharging and receiving services to coordinate patient care responsibilities. This, in turn, can force clinicians to selectively implement orders and to make ad hoc decisions about which orders apply in one setting and which ones should be held until the patient is moved elsewhere. Noted one study participant:

> There is a major problem with confusion over whether it is the floor accepting the patient or the ER transferring. The difference is who is responsible.

Similarly:

> Patients sometimes don't get out of recovery for half a day or a day because there is no bed to move them to. With paper orders, you [write] the orders for the intensive care unit (ICU) or the floor . . . and the nurses [are able to] carry out the floor care or ICU orders [in the recovery area] . . . this doesn't work in CPOE.

CPOE systems often enforce very rigid scheduling, such that the timing of tests or medication administration cannot be altered. These systems typically tie scheduling to the time the order was processed, not the time it was actually executed, making it difficult—if not impossible—for staff to alter the timing information to match reality. This can cause critical delays in medication administration:

One problem was that the start time in our system doesn't mean the time the medication is first administered—it means the time the order becomes active and then the administration times are automatically calculated based on that. [A physician ordered] a q12 medication. The first scheduled administration time was about 11 hours later, so the patient's post-transplant medication was delayed about 11 hours beyond when the physician wanted [it].

In addition, once an initial dose of a medication is given, CPOE systems can make it difficult for clinical staff to alter the timing of subsequent doses, such as when patients must have medications withheld prior to surgery, or when they are absent from the nursing unit for tests or procedures when medications are due. Even in systems where medication dosage times can be changed, often the system cannot automatically reschedule subsequent doses after this modification, requiring staff to alter each of the remaining dosage times manually to match the new, correct administration schedule.

Alterations in work pace, sequence, and dynamics represent changes that emerge primarily from the difficulties inherent in computationally modeling the nonlinear, iterative, ad hoc, interruption- and exception-driven activities that comprise clinical care.[4,6] These effects are compounded by the fact that CPOE often computerizes only one segment of clinical care: the ordering process. This can dramatically affect the care delivery process as a whole, as patterns of communication, cooperation, and collaborative work must shift to accommodate the technology.[9] It is not surprising that the National Health Policy forum has reported that clinician productivity can drop approximately 20% within the first 3 months of a new CPOE implementation,[3] though other studies[10] have indicated that productivity often improves over time as users gain proficiency with the system.

Clearly, there is much work to be done to improve CPOE system design, particularly with regard to restructuring of awkward implementations of clinical tasks, improving interoperability with and access to other clinical information systems, and gaining a better understanding of how users circumvent the system to get their work done. As our understanding of the sociotechnical aspects of clinical work and CPOE systems evolves, we expect substantial improvements in the fit between clinical work processes and the computer technologies designed to support them.

3.3.3 CPOE Design Does Not Include Other Clinical Staff

In the study conducted by the authors, nonphysician staff (especially nurses) voiced strong and repeated concerns about not being included in the process of CPOE development and integration. For example, nurses find that they have to train physicians to enter orders covering activities that until now have been part of standard nursing practice, meaning that nurses may be prevented from carrying out certain routine activities until the order is entered into the system:

> This is not a nursing system . . . the nurses are just saying, "Give me a template nurses can use. Give me standard order sets I can sign off with a single review. Get the standard nursing orders into the doctor's order templates, so we don't have to remind them to write an order for something like drawing arterial blood gases every 8 hours unless the patient condition changes." This is just standard [nursing] practice.

Many staff found it especially bothersome to be alerted to issues that were not applicable in the current setting for the particular staff member. This is especially common when the organization implements system-wide alerts without paying careful attention to their usefulness in specific locations. For example, some drug–drug interaction alerts may be highly desirable in one context but not in another. One intensive care nurse observed:

> Some alerts can be really annoying—particularly the ones warning against prescribing heparin and aspirin. These are CCU patients; the system should know we are going to give these two meds together on this floor and quit warning us about them. We get that warning over and over.

In a paper-based system, orders are accompanied by the repeated review and helpful input by various medical professionals. This implements a system of checks and balances that can optimize order-processing activities. In the course of standardizing practice and streamlining steps, the elimination of these fundamental interactions within the CPOE ordering process can directly affect workflow. For example, without order review by ward secretaries, nurses, pharmacists, and others, errors and inconsistencies may enter and propagate throughout the system. Staff must develop workarounds to accommodate nonauthorized staff, such as when a resident must enter a medical student's orders because the system does not authorize the student to do so. In these ways, some of the clarification provided by constant order oversight is lost when new CPOE processes are substituted for historical ones, thereby creating subtle but important mismatches between the system and the people who interact with it.

Healthcare delivery is a complex activity system that requires the expertise of various professionals whose respective skills are interrelated and inseparable. Indeed, this distribution of work enhances the robustness of the system.[11] However, current CPOE systems do not always accommodate or model the work needs of all levels of clinical personnel. In fact, many CPOE systems seem to provide support for only the primary order-entry person's work activities. Although physicians bear the legal responsibility for order entry and have the expertise needed for the decision making that is required, the ordering process requires many different levels of healthcare personnel to complete it. For these reasons, it is paramount that the roles of nursing, clerical, pharmacy, and all other ancillary staff be carefully considered when CPOE systems are designed and implemented (and, indeed, throughout the lifetime of the software) if workflow is to proceed with minimal disruption. Because the sociotechnical approach to system study includes all components of the system, such studies should help us identify omissions in CPOE functionality and clinical coverage.

3.3.4 CPOE Reduces Situation Awareness

Dourish and Bellotti define situation awareness as "the understanding of the activities of others which provides a context for your own activity."[8] Collaboration understandably improves when people can develop and maintain awareness of what is going on around them.[6] We notice that CPOE systems, because they allow orders to be entered at any time by providers located outside the hospital, can contribute to loss of situation awareness. One very clear example illustrates this point:

> The model of care there was just different than what we had seen other places. It was not at all unusual in the paper world to have two or three people generate orders very close to each other, but the common thing they had was a paper or a sheet. In the emergency department, [there] was literally a different workstation about every two inches down there. We had a lot more instances of within 30 seconds of each other, two—sometimes three—providers would enter the same order at approximately the same time, and so it really forced us to go back and do more education on being careful to look and see what [orders are] active before you enter a new order.

CPOE can also dramatically alter where, how, and when orders are entered because it allows physicians to enter orders from off-site locations or at times distant from the actual patient interaction. Although this is a great benefit in terms of flexibility for clinicians, it can result in changes in the timing of order placement and thus directly affect patient care:

> Residents making ICU rounds in the past wrote the orders, handed them to the nurse. They wrote them right away and they did them in the ICU. In our system, now they make the rounds, they leave, they go to conference, they go to lunch, they go see their new admission, and then they pull out a piece of paper on which they scribbled some notes to themselves and say, "What was I going to order on that guy in the ICU?" about 4 hours later, and they do it from a remote site so that when they enter it, the patient status may have changed.

Unless the CPOE system is integrated with other clinical information system components that can provide a complete picture of the state of the patient at the time the orders are actually entered, the provider may be operating in an information vacuum.

Interesting situation awareness issues can emerge when providers from different clinical services use CPOE to enter orders simultaneously on the same patient. The orders might appear to conflict, when, in fact, they do not:

> I was sitting there in the ICU looking at my patient and all of the sudden, boom, an order for dopamine shows up. I didn't write that . . . and I look at it . . . and turned out that it was written by the anesthesiologist getting ready for the case tomorrow. So I was seeing all of the pre-op medicines . . . a good thing, right? Except it surprised me. I've never seen those orders before, and [the patient] looked like he didn't need dopamine to me, so I just cancelled the order.

CPOE systems can vastly improve situation awareness through functionality that is integrated with other systems and subsequently displays information derived from these differing sources in a single location. In addition, clinical decision support tools can alert clinicians to potential problems that might otherwise go unnoticed (e.g., drug–drug or drug–allergy interactions). Because CPOE can improve practice standardization (e.g., through order sets or codification of procedures), such a system can provide a level of consistency in practice that can enhance situation awareness, as users learn to "expect" certain CPOE behavior, and adapt to CPOE processes.

At the same time, CPOE can contribute to a general loss of situation awareness when it changes the pattern, style, and timing of provider interactions. As mentioned earlier, collaborative work processes are essential to the effective delivery of health care. Although CPOE systems are intended for multiple providers, designers of these systems do not appear to consider (or have not yet been able to implement) the degree to which iterative and interactive communication among the various players is essential to promote and support situation awareness in medical work and decision making.[12] Such awareness is vitally important to ensure effective performance in any complex and dynamic environment.[13] Without careful design to facilitate multiple-provider communication, the computerization of health records can serve to isolate users from one another, depriving them of the benefit of coworkers' understanding of and insights into the clinical situation. Continued research into the organization and display of information, including the development of new techniques to help alert clinicians or help them locate and interact with other clinicians using the system (e.g., hands-free, wireless Internet telephony devices with speaker-independent speech recognition capabilities), may serve to improve situation awareness when using CPOE systems.

3.3.5 Ineffective Implementation of Policy and Procedures Can Be Highlighted by CPOE

CPOE systems help to formalize organizational policies and procedures.[9] In many cases, actual practice does not match this rigid "letter of the law," meaning that the CPOE system may introduce a significant amount of extra work (perceived or real):

> We found that [obstetrics] was one of the most complex places in the hospital because patients were going from the screening room to the pod room to a labor room to the delivery room to postpartum and each of those [is] a different level of care and so orders need to be rewritten. Although nurses are very good about blending the orders as need be from one [level] to the other, the computer isn't nearly as flexible.

In addition, difficulties can arise when standards are difficult to interpret or implement, as when one clinician initiates patient care that must be monitored by other specialists:

> Some orders [are] written by certain specialists like anesthesiologist [for] epidurals. No one wants to rewrite those orders. So how should those [orders] traverse the levels of care when the epidural catheter moves with the patient?

In such cases, CPOE can complicate an already difficult issue.

Some CPOE systems do not allow certain clinical providers to enter orders, making it necessary for a provider to enter orders for another person, such as a physician entering orders for a respiratory therapist. Because these providers often do not round together, the physician may have difficulty verifying that the correct information is entered:

> You used to have the respiratory therapists round with the doctors . . . The RT would then spout off the SpO_2, PEEP, tidal volume or whatever, then the doctor would write the orders. We have a big change now. The doctors have to write the orders into the computer, and the RT isn't usually there beside them, so they are either asking the nurses what the settings are, or they have to go find RT or go check the ventilator themselves. This has had a big effect on the doctors.

While CPOE can be a highly effective and efficient tool for implementing organizational policies and procedures, using CPOE in this manner can bias workflow design toward an organizational perspective, one that emphasizes an explicit view of work: "those things that are documented, visible, and articulable [sic],"[14] such as procedures and methods. This view fails to acknowledge the more tacit aspects of work processes—those activities carried out in everyday practice, which rely on human ingenuity and depend on rules of thumb or individual judgment for synthesis and completion,[14] or what Berg called "the user perspective."[1] As a result, rote implementation of policy or procedure can highlight pronounced differences between organizational intention and provider practice, leading to the adoption of system workarounds by clinicians who are struggling to use a system that does not fully support their work. Problems that exist in a paper-based environment can be compounded by rules that are either irrelevant or counterproductive in a computerized environment.

Any implementation of organizational rules or directives should be undertaken only after careful assessment of the impact of such changes on actual clinical work, to determine whether these rules can be *practically* integrated into workflow. In addition, care must be taken to assure that work practices mandated in CPOE systems are actually formally (e.g., legally) required, as opposed to representing "the way things have always been done." CPOE can make the process of implementing organizational changes seem easy. Without sufficient attention being paid to the unintended adverse consequences of these changes on workflow practices, however, problems can occur. It is, therefore, imperative that any organizational mandates implemented through CPOE are rigorously tested using real-time scenarios to assure that these requirements not only make practical sense but also do not negatively impact workflow.[15]

3.4 Conclusion

The introduction of CPOE into the healthcare environment has a dramatic effect on clinical workflow. CPOE systems are tools intended to support and improve the delivery of care, but they are not solutions for all problems related to clinical practice. The sociotechnical approach suggests that technology and workflow should "fit" together and that improving this fit is the goal of continued study. Over time, each side of the equation will need to change, requiring the other to adapt in turn. By observing and interviewing clinicians, we found that CPOE systems can alter work practices in several consistent ways, regardless of the site studied.

In our study, we identified five repeating themes:

1. Significant interface and ergonomic problems can arise as computers are placed in environments not initially designed to accommodate them and are used by people unfamiliar with the technology.
2. CPOE can alter the pacing, sequencing, and dynamics of work patterns.
3. Despite the fact that these systems are ostensibly *provider* systems, they remain— at least in the sites we studied—predominantly *physician* systems, such that the workflow needs of nonphysician personnel are often poorly implemented or ignored altogether.
4. CPOE can affect situation awareness for providers, such that clinicians cannot guarantee that they are acting on full or complete information at all times.
5. CPOE can be leveraged to poorly implement organizational policies and procedures, creating extra work or slowing down current work processes for providers.

We must take care to continually improve these systems if they are to fit seamlessly into the clinical workflow. As we identify how, when, and where workflow mismatches arise, we gain more insight and can design better tools for building more sophisticated systems. As CPOE systems evolve, ongoing care must be taken to reduce or resolve the many unintended adverse effects that these systems can have on clinical workflow. The five kinds of unintended and unanticipated consequences related to workflow with CPOE can be mitigated by iteratively altering both clinical workflow and the CPOE system until a comfortable and optimized fit is achieved.

Questions to ask regarding workflow issues in relation to CPOE include:

1. What is the ratio of workstations to staffed beds for the inpatient setting (sorted by ICU and acute care)?
2. What is the ratio of workstations to clinical exam rooms now in use for the outpatient setting?
3. What is the distribution of computing capability among the majority of clinical workstations (e.g., speed of CPU, amount of RAM, size of hard disk, size of monitors)? If the system were running Citrix, how would this change things?
4. How many days behind is the organization in upgrading to the CIS vendor's most recent product release?
5. What percentage of the patients' medical records exists in an easily retrievable electronic format (e.g., laboratory test results, radiology reports, images, surgical reports, outpatient progress notes)?
6. Can clinicians access all of their patients' data from any location within the hospital as well as remotely (home or office)?
7. What is the record of system response time measured (to the tenth of a second) from the users' perspective (includes delays resulting from database, network, application, and workstation)?

8. What is the record of network performance (e.g., percentage of capacity, transmission speed)?

9. What is the record of system uptime (or downtime)—measured to the minute— of the top five clinical applications (should include both planned and unplanned downtime of all aspects of the system that affect users, such as databases, the network, applications, application interfaces, and workstations)?

10. Is there is an enterprise-wide authentication system so that users can log on once for each session and access any system, function, or database to which they have access privileges?

11. What is the percentage of active (responsible for some aspect of patient care) clinicians (e.g., physicians, RNs, physician assistants) who log in to some portion of the CIS infrastructure on a weekly basis?

12. What percentage of orders (e.g., medications, laboratory, radiology) is entered by the authorizing provider who made the decision? (This may be a physician, physician assistant, or nurse practitioner.)

13. What is the percentage of all orders requiring cosigning that are not cosigned within 48 hours? (This percentage should be monitored and reviewed periodically.)

14. What percentage of active clinical users log in to the CIS? (This percentage should be monitored and reviewed periodically.)

15. What is the percentage of order sets for top 100 admitting diagnoses for the organization that are available?

16. Is there periodic (quarterly, semi-annual, annual) monitoring and review of the percentage of all clinical alerts and reminders that actually fire?

17. Does a notification mechanism or procedure exist for all new clinical orders entered into the system?

18. Are there two-way computer interfaces between the CPOE system and the clinical laboratory, pharmacy, radiology system, and dietary system?

19. What percentage of orders is entered by physicians or the person responsible for making the clinical decision? (This percentage should be monitored for all major order types.)

20. What percentage of all orders is entered as "stat" or "ASAP"? (This percentage should be monitored and reviewed periodically.)

21. What percentage of pharmacy orders is modified (edited, completed, altered, changed, and so on) by pharmacists?

22. What percentage of radiology orders is modified (edited, completed, altered, changed, and so on) by radiology technicians?

23. What percentage of all orders of each type (e.g., medications, laboratory, radiology, nutrition) is cancelled? (This percentage should be monitored.)

24. Does the system have the ability to print inpatient rounds reports (at least 24 hours of data)?

25. Does the system have the ability to print pre-visit patient summaries in the outpatient setting?

26. Is there an institutional-level policy and computer-based procedure for handling the paper portions of the medical record (scanning)?

27. Does the organization have a means of identifying one physician who is responsible for each patient at each point in time? Some organizations identify both a clinician responsible for "responding" to an event and a clinician "responsible" for the patient. Such a link is needed if the CIS needs to contact a clinician regarding a critically abnormal test result, for example.

3.5 References

1. Berg M. Patient care information systems and health care work: a sociotechnical approach. *Int J Med Informatics.* 1999;55(2):87–101.

2. Crabtree BF, Miller, William L. *Doing Qualitative Research.* 2nd ed. Thousand Oaks, CA: Sage; 1999.

3. Campbell EM, Sittig DF, Ash J, Guappone K, Dykstra R. Types of unintended consequences related to computerized order entry. *J Am Med Inform Assoc.* 2006;13:547–556.

4. Berg M. Medical work and the computer-based patient record: a sociological perspective. *Meth Inform Med.* 1998;37(3):294–301.

5. Reddy MC, McDonald DW, Pratt W, Shabot MM. Technology, work, and information flows: lessons from the implementation of a wireless alert pager system. *J Biomed Informatics.* 2004;38(2005):229–238.

6. Kaplan B. Evaluating informatics applications: some alternative approaches: theory, social interactionism, and call for methodological pluralism. *Int J Med Informatics.* 2001;64(1):39–56.

7. Strauss A, Corbin J. *Basics of Qualitative Research: Techniques and Procedures for Developing Grounded Theory.* Thousand Oaks, CA: Sage; 1998.

8. Dourish P, Bellotti V. Awareness and coordination in shared work spaces. In: *ACM Conference on Computer-Supported Cooperative Work* (CWSW '92). Toronto, Canada: ACM Press; 1992:107–114.

9. Kuperman GJ, Gibson RF. Computer physician order entry: benefits, costs, and issues. *Ann Intern Med.* 2003;139:31–39.

10. Overhage JM, Perkins S, Tierney WM, McDonald CJ. Controlled trial of direct physician order entry: effects on physicians' time utilization in ambulatory primary care internal medicine practices. *J Am Med Inform Assoc.* 2001;8(4):361–371.

11. Hazlehurst B, McMullen C, Gorman PN, Sittig DF. How the ICU follows orders: care delivery as a complex activity system. *Proc AMIA Annu Symposium.* 2003;284–288.

12. Gennaria J, Weng C, Benedetti J, McDonald D. Asynchronous communication among clinical researchers: a study for systems design. *Int J Med Informatics.* 2005;74:797–807.

13. Endsley M. Toward a theory of situation awareness in dynamic systems. *Human Factors.* 1995;37(1):32–64.
14. Sachs P. Transforming work: collaboration, learning, and design. *Commun ACM.* 1995;38(9):36–44.
15. Sittig D, Ash J, Zhang J, Osheroff J, Shabot M. Lessons from "Unexpected increased mortality after implementation of a commercially sold computerized physician order entry system." *Pediatrics.* 2006;118(2):797–801.

4 Emotional Aspects of Clinical Information System Use

Based on: Sittig DF, Krall M, Kaalaas-Sittig J, Ash JS. Emotional aspects of computer-based provider order entry: a qualitative study. *J Am Med Inform Assoc.* 2005;12(5):561–567.

4.1 Key Points

- Computer-based provider order entry (CPOE) systems are implemented to increase both efficiency and accuracy in health care, but these systems often spur myriad emotions among healthcare personnel.
- In the study discussed in this book, the implementation and use of CPOE systems provoked examples of positive, negative, and neutral emotions.
- These systems and the implementation process itself often inspire intense emotions. Negative emotional responses were by far the most prevalent in the study.
- Designing and implementing CPOE systems is difficult.
- If designers and implementers fail to recognize that various CPOE features and implementation strategies can increase clinicians' negative emotions, then the systems may fail to become a routine part of the clinical care delivery process.
- Some of these problems might be alleviated by designing positive feedback mechanisms for both the systems and the organizations.

4.2 Introduction

Emotions are mental states that arise spontaneously, rather than through conscious effort (see the work of Morelos-Borja[1] for an overview of emotions and their relationship to technology). Changes in underlying physiology, facial expression, or even actions often accompany emotions. A specific event or series of events that either cause the person to succeed or fail in reaching his or her goal(s) triggers most emotions. One's emotions and resulting moods reflect on a person's ability to attend to complex physical and cognitive tasks.[2]

While emotions of all types are interesting to study, in this book we were interested in those emotions associated with decreases in cognitive and physical abilities. CPOE seems to elicit such emotions. For example, Ash et al.[3] identified many examples of emotionally loaded words related to CPOE, either uttered by house staff

Table 4-1   The Taxonomy of Emotionally Related Terms Used to Classify the CPOE-Related Quotes

Positive Terms	Neutral Terms	Negative Terms
Love: lust, attraction, adoration	**Sleepy:** drowsy, listless	**Shame:** humiliated, embarrassment, shy
Liking: admiration, friendly, affection, excellent	**Apathetic:** apathy, boredom	**Sadness:** melancholy, wistful, guilt, sorrow
Contentment: harmony, peaceful	**Contemplative:** calm, dreamy, strange	**Pain:** agony
Happiness: amusement, hope, gaiety, joy	**Arousal:** electric	**Anxiety:** worried, nervous, tension, upset
Pride: satisfaction, smug	**Interest:** alert, fascination, curious	**Fear:** afraid, dread, terror, shock
	Surprise: amazement	**Anger:** mad, annoyed, indignant, cross
	Understanding: confusion, uncertain, confidence	**Hostility:** hatred, revenge, defiant, spite, dislike, distrust, bitterness
		Disgust: contempt, distaste

Adapted from: Storm C, Storm TA. Taxonomic study of the vocabulary of emotions. *J Pers Soc Psychol.* 1987;53:805–816.

or used to describe house staff, such as, "The surgeons have always used it, and the internists have sort of been laggards" or "They are pretty intolerant if anything isn't just the way they want it."

There is no widely accepted taxonomy of emotions.[4] In an effort to better study emotions by analyzing textual documents, Storm and Storm[5] developed a hierarchical taxonomy of semantically homogeneous terms that are associated with specific emotions. They further grouped these terms into negative, positive, and neutral emotions. After careful review, we believe that their taxonomy is sufficient to begin illustrating the concepts that we are after (see Table 4-1).

4.3 Ways in Which Emotions Lead to Unintended Consequences

The first section describes the positively focused emotions that the CPOE systems provoked in the study described in this book. The second section describes those emotions that are more neutral, and the final section illustrates the myriad negative emotions that these systems may generate.

4.3.1 Positive Emotions

Our analysis found very few instances in which people interviewed or observed mentioned or expressed any sort of positive emotion resulting from the CPOE system. In this particular case, the "positive" emotion is more a lack of a negative emotion than a true positive experience. For example, one clinician stated:

> I would look positively on that [medication interaction alerts] because there are more and more [medication] interactions and so many choices of medications now, that I think it doesn't hurt to have reminders like that.

Happiness/Enjoyment

Interestingly, the aspects of the systems that tended to provoke these "good feelings" were, for the most part, unintentional or serendipitous on the part of system designers. For example, in one particular system, health maintenance reminders are indicated by a small text box in the upper-right corner of the screen; this text box is highlighted if a reminder exists. Once the user has satisfied the alert by ordering the appropriate test or documenting an extenuating circumstance, the highlighted text returns to normal. One user said:

> It's a big deal for me to have those lights unlit. It is a reward for me to have that light unlit. It's like guessing what the health maintenance is. I mean, you have to have these little games you play all day. It's like, guess what the health maintenance thing is? Pop it off. Your patient calls, pop it off your messages. It's like a little reward system.

Along those same lines, another user chimed in:

> I do miss the little alert when you empty it [the in-basket]. There used to be a little box that would pop up that would say, "Your in-box is empty," and a little sound went off. I used to like that because it was such a victory. You guys took that one away [group laughter].

Liking/Attachment

While not a unique attribute of CPOE, clinicians like the fact that it is no longer necessary for them to have the patient's chart in their possession to write an order with a CPOE system. Said one clinician, "One of the beauties of not having a paper chart is you can do it anywhere, and you can do it on the fly. And so that's actually one of the great bonuses of this process."

Pride/Honor

Recognition by one's peers is one of the greatest honors that a person can achieve. By creating useful order sets, a clinician in an organization where personal order sets are shared can make a name for himself or herself within the organization. As one clinician stated:

> [E]fficiency was the first thing, but in fact, I think one of the things that happened, that was a side benefit, is that the ownership [of the order sets] became an issue and so now you can construct your own personal order set. And that became yours—and you could mold it however you wanted and it would be, you know, your hallmark.

Contentment/Satisfaction

No one familiar with the current state-of-the-art clinical computing systems and the difficult CPOE implementation process would ever expect the initial version of a system to be perfect. In contrast, by responding to user criticisms, system administrators have an excellent opportunity to win over clinicians. Said one system administrator, "[The relationship with clinicians is] much improved now. And there's aspects to [CPOE] that actually make things easier than they used to be."

4.3.2 Neutral Emotions

Apathetic/Indifferent

CPOE systems are not as complex as the human emotions and interactions that surround them in healthcare institutions. While these systems often provoke negative emotions in junior members of the academic healthcare team, senior members of the team often show indifference toward them. Whether feigned or real, this indifference may increase the negative feelings of the junior members of the team because the senior members often are not required to use the system. For example, a senior clinician said:

> I think . . . the actual entry of orders is not really as important in my mind as the thought process of what the plan is and what we want to do. And if we all, in a teaching environment, we all make that decision together, I don't care who the hell puts the orders in.

Contemplative/Reflective

While stress of any kind can bring out the worst in some individuals, others often rise to the occasion when faced with extra pressure. These extraordinary individuals' efforts are often recognized only long after the situation has been resolved:

> [T]here were several individuals who devoted . . . hours of their time, days, months of their time trying to work within their specialty. And these people sort of rose to the top, became identified by [Joe], [George], me, others as being interested in trying to work through the problem—and they . . . ended up shouldering the burden simply because they're that type of person in trying to work through this with their colleagues: "Now what do we want?" "Here's how our order sets are going to look," and that sort of thing.

Interest/Curiosity

CPOE implementations often inspire intense emotions in members of the healthcare team. The following quote illustrates how one clinician described the three groups of clinicians (interested, apathetic, and saboteurs) at his institution:

> [P]robably people fell into three groups: ones who realized and admitted to themselves that this was going happen regardless of what they did [e.g., were interested and helped out with the process]; those who were ambivalent either way and knew that it was a real pain in their life but didn't care enough to take action one way or the other; and then the group who were reactive and wanted to try to dump the system as the solution rather than work through it.

Understanding/Tolerance

Often the most one can expect from a group of disgruntled employees is that they develop some level of understanding, or tolerance, of the system. Luckily, in one particular instance, that is just what the administration received:

> [W]ith time, everybody worked it out [clinicians and administration agreed to continue with CPOE implementation]—that it [a strike] would never be perceived as a positive sort of thing from a public relations point of view and that the administration wouldn't [budge] if you have it on that issue [patient safety], and that things would be worse rather than better [if they struck with, rather than fixed, the CPOE system].

4.3.3 Negative Emotions

Negative emotional responses to the CPOE systems were by far the most prevalent in all of the observations.

Shame/Guilt

Many times users are upset by the position in the workflow in which alerts arise. For example, one respondent said, "I feel guilty if I click it 'no' [when asked, "Do you want to document your actions to satisfy this alert?"], because it's almost as though I haven't dealt with it, but I always do deal with it." Another stated:

> We take these alerts personally, and they're like a slam. It's like, "Well, they [the patient] just got here and I haven't even had a chance to do anything [e.g., enter an order for a specific preventive screening procedure], and I'm getting the ALERTS!"

Anger/Annoyance

CPOE system designers think they are helping clinicians take the correct actions and hope that the alerts and reminders that pop up to interrupt the clinician's workflow will help them "learn" to perform the correct action. Unfortunately, these little "learning opportunities" do not always have the intended effect, as McDonald[6] showed more than 25 years ago. As one clinician said:

> It just makes you angry. . . . even if you are ordering what it recommends, you still get the little message. Yeah, that's maybe the most annoying.

Sadness/Melancholy

Some respondents reflected that life is short and that their time could have been spent more productively. In this case, a sense of sadness often ensues:

> [F]or me personally, I gave up a tremendous amount of personal time for my family [to work on the CPOE system]. I gave up a lot of my time that I probably should have been devoting to my research program to deal with the [CPOE] issues. . . . also a lot of educational time, going to meetings, going to seminars, that sort of thing. So from a personal sense, there was a quite a bit of sacrifice in that sense.

Hostility/Animosity

Hostility was one of the most commonly expressed emotions throughout all interviews and focus groups. A work stoppage, even if only threatened, represents a level of hostility rarely seen in, or even around, healthcare institutions:

> But they . . . happened like a hurricane [e.g., rallies of hostile residents] within the space of 2 or 3 days. All of a sudden something would happen, stir the pot, [and] everybody would decide, "Okay, we've had enough and this is the time [to strike]."

Disgust/Loathing

Disgust often results from situations in which the person has little or no control over the situation at hand. These feelings of disempowerment were widespread throughout the data sets in the study under discussion:

> Everybody felt like the university had gotten the short end of the stick, they'd . . . got a bad product. And then to make it [worse], we couldn't dump it and it [was a] $6 million investment—whatever it was, I don't know how much the investment was— it was a lot of money and you couldn't just dump it into a different vendor

4.4 Discussion

The practice of medicine is filled with enormous stressors, including an incredible workload[7] that is often accompanied by a lack of sleep,[8] the threat of lawsuits, feelings of loss of control over the workload and/or schedule,[9] task interruptions,[10] and patients with complex medical problems.[11] With all these stressors capable of causing negative emotions, CPOE system designers and implementers must be especially careful not to cause a nonlinear negative effect on performance (e.g., failure to complete a task or an erroneous action)[12] by popping up yet another irrelevant alert[13] or implementing a new CPOE system[14] on an underpowered and slow computer system.[15] Likewise, clinical system designers and healthcare administrators should be careful to listen and respond in a positive manner to clinical users' concerns. At the same time, clinicians need to be educated about the potential long-term benefits of CPOE and remain open to the possibility of changes in their workflow.

CPOE systems with clinical decision support capabilities frequently focus on alerting or reminding clinicians when they have forgotten to do something or when they have done something wrong rather than trying to educate or help them to do a better job, by making "the right thing to do the easiest thing to do." It is little wonder that so many of the emotions exhibited by clinicians involved with CPOE implementation—or any clinical information system, for that matter—are negative.[16,17] Interestingly, many of the positive emotional responses that we observed in our study resulted directly from the following sources:

- Positive feedback from other users (e.g., the clinician's order set becomes a "hallmark")

- Evidence that either system designers or healthcare administrators listened to their concerns and made positive changes (e.g., "System is much improved now" or "Clinicians and administrators worked it out")

- Aspects of the systems that provided some sort of positive feedback to clinicians (e.g., "lights unlit" or "little sound")

Likewise, many of the negative emotions that we recorded resulted from either negative feedback from the system or from healthcare or information technology administration personnel. Based on these findings and the recognition that clinical information systems users already lead stressful lives, system designers might consider designing systems that provide more positive—rather than punitive—feedback.[18]

4.4.1 An Idea for Improving Clinical Information Systems by Providing Positive Feedback

While CPOE system designers and implementers can never hope to remove the myriad stressors routinely faced by clinicians, it might be possible for them to develop new system features that provide positive feedback to clinicians—although we must be careful not to further disrupt the clinicians' work while doing so. For example, the ability for each clinician to select first whether he or she would like to receive positive feedback and, if so, to what degree, might be helpful. If a clinician chose to participate in the positive feedback program, he or she might receive reinforcement like that in the following examples:

- The system might pop up a small congratulatory reminder whenever a clinician enters 100 medication orders in a row that are on the formulary.

- A small geometric figure could be placed in an unobtrusive location on the screen that has pieces added to it, or is colored in, as the clinicians work with the system as designed.

- A congratulatory message could be sent to clinicians by the system when they have completed all their patient encounters within 3 days of the event over a 1-month period.

- A pleasant audio sound might be generated when clinicians perform a series of tasks correctly.

- Clinicians could receive a $50 gift certificate whenever 90% of their patients meet their health maintenance targets in any given month.

- Clinicians who have overridden a particular alert more than 90% of the time and more than 10 times in the past 3 months could be presented with the following options: Stop receiving this alert because . . . [enter reason here] or Continue receiving this alert.

- We could allow physicians to adjust the level of positive feedback that they receive so that they would not receive more disruptions than they want.

If clinicians received such positive reinforcements or simply felt more in control of their workflow, they might react differently to the various alerts and reminders that are currently in place. For example, rather than feeling as if the alerts are "punishing" them, they might begin to believe that these alerts are "helping" them reach their objectives.

4.5 Conclusion

The analysis identified all of the emotions listed in Table 4-1; negative emotional responses to the CPOE systems were by far the most prevalent in all the observations. While we interviewed or observed more than 50 people, the number of instances of any positive emotion resulting from the CPOE system was very small. Based on our analysis, the most one can expect from overworked employees who are asked to undergo a major change in their workflow is that they develop some level of understanding or tolerance for the system. These "neutral" emotions were much more common than positive ones in our study but far less common than the negative emotions (e.g., negative emotions >> neutral emotions >> positive emotions).

Designing, implementing, and maintaining CPOE systems is difficult. The implementation and subsequent use of these systems inspire intense emotions in nearly everyone involved. If system designers fail to recognize that various CPOE features and implementation strategies can increase clinicians' negative emotions, then this omission may increase the likelihood that the system implementation will fail or that the system will never be routinely used by clinicians. Increasing the positive feedback associated with these systems might alleviate some of these problems, although this hypothesis remains to be tested.

Questions to ask regarding staff emotions issues in relation to CPOE include the following:

1. Are regularly scheduled (at least monthly) clinical information system oversight meetings held?
2. Is 24 × 7 × 365 telephone help-desk support available for all clinicians?
3. What percentage of calls for help are answered on the first ring?
4. What percentage of calls for help are completed/helped on the first call?
5. What percentage of regularly scheduled clinical information system oversight meetings have at least one practicing physician (e.g., who uses the CIS to review and record patient information) and one nursing representative present?
6. Are there incentives in place to encourage physician involvement in the process?
7. Are clinicians paid for the time they are involved with clinical information system activities (e.g., meetings, training, design, development, testing)?
8. How are the benefits of clinical information system usage explained and communicated to members of the organization?
9. What percentage of orders are canceled within 5 minutes of their entry?

10. What is the mean time from initial query to order selection? This parameter measures how long it takes the average clinician to find what he or she is looking for and provides a gross measure of system usability.

11. Are there employees with extensive clinical experience (e.g., RN, PharmD, MD, lab technician) in the Information Technology department?

12. Does the organization pride itself on being a leader in innovation, among the first to adopt new practices and technology, or does it prefer to wait until there is clear scientific evidence of benefit before jumping on board?

13. To what extent is clinical care standardized throughout the organization?

14. Does the organization have a strong track record of successful (e.g., on-budget, on-time, achieved goals) clinical system (with or without information systems) improvement projects?

15. Does the organizational culture value constructive feedback, changes made for quality improvement, and continuous learning—kept in balance by leadership that can tell the difference between clinicians' requests for "what would be nice" versus "what is essential or critical for success"?

16. Can the organization afford the temporary productivity losses that accompany most large CIS implementation projects?

17. Have adequate funds been dedicated solely for CPOE, with the ability to commit additional funds quickly for good (unanticipated) causes?

4.6 References

1. Morelos-Borja H. *Partner-Finder: a Framework to Study Peer Collaboration in Web-Based Education*. Doctoral Thesis. UMI Order Number: AAI 9939122; 1999; University of Central Florida.

2. Picard RW. *Affective Computing*. Cambridge, MA: MIT Press; 1997.

3. Ash JS, Gorman PN, Lavelle M, et al. Perceptions of physician order entry: results of a cross-site qualitative study. *Methods Inf Med.* 2003;42:313–323.

4. Mozziconacci S. Prosody and emotions. Presentation at Speech Prosody 2002, Aix-en-Provence, France, April 11–13, 2002. Available at: http://www.isca-speech.org/archive/sp2002/sp02_001.pdf. Accessed August 16, 2009.

5. Storm C, Storm TA. Taxonomic study of the vocabulary of emotions. *J Pers Soc Psychol.* 1987;53:805–816.

6. McDonald CJ. Protocol-based computer reminders, the quality of care, and the non-perfectibility of man. *N Engl J Med.* 1976;295:1351–1355.

7. Appleton K, House A, Dowell A. A survey of job satisfaction, sources of stress, and psychological symptoms among general practitioners in Leeds. *Br J Gen Pract.* 1998;48:1059–1063.

8. Papp KK, Stoller EP, Sage P, et al. The effects of sleep loss and fatigue on resident-physicians: a multi-institutional, mixed-method study. *Acad Med.* 2004;79:394–406.

9. Deckard G, Meterko M, Field D. Physician burnout: an examination of personal, professional, and organizational relationships. *Med Care.* 1994;32:745–754.

10. Chisholm CD, Collison EK, Nelson DR, Cordell WH. Emergency department workplace interruptions: are emergency physicians "interrupt-driven" and "multi-tasking"? *Acad Emerg Med.* 2000;7:1239–1243.

11. Wetterneck TB, Linzer M, McMurray JE, et al. Society of General Internal Medicine Career Satisfaction Study Group. Worklife and satisfaction of general internists. *Arch Intern Med.* 2002;162:649–656.

12. Kolich M, Wong-Reiger D. Emotional stress and information processing ability in the context of accident causation. *Int J Indust Ergon.* 1999;24:591–602.

13. Weingart SN, Toth M, Sands DZ, Aronson MD, Davis RB, Phillips RS. Physicians' decisions to override computerized drug alerts in primary care. *Arch Intern Med.* 2003;163:2625–2631.

14. Shu K, Boyle D, Spurr C, et al. Comparison of time spent writing orders on paper with computerized physician order entry. *Medinfo.* 2001;10:1207–1211.

15. Yamauchi K, Ikeda M, Suzuki Y, Asai M, Toyama K, Hayashi E. Evaluation of the order-entry system by end users: a step to the new hospital information system. *Nagoya J Med Sci.* 1994;57:19–24.

16. Darbyshire P. "Rage against the machine?": nurses' and midwives' experiences of using computerized patient information systems for clinical information. *J Clin Nurs.* 2004;13:17–25.

17. Lundberg U, Melin B, Evans GW, Holmberg L. Physiological deactivation after two contrasting tasks at a video display terminal: learning vs. repetitive data entry. *Ergonomics.* 1993;36:601–611.

18. Eisenberger R, Cameron J. Detrimental effects of reward: reality or myth? *Am Psychol.* 1996;51:1153–1166.

5 Shifts in Power, Control, and Autonomy

Based on: Ash JS, Sittig DF, Campbell E, Guappone K, Dykstra RH. An unintended consequence of CPOE implementation: shifts in power, control, and autonomy. *AMIA Annu Symp Proc.* 2006;11–15.

5.1 Key Points

- A multidisciplinary team found that CPOE enables shifts in power related to work redistribution and safety initiatives and causes a perceived loss of control and autonomy by clinicians.
- With recognition of the extent of these shifts, clinicians can anticipate them and will no longer be surprised by them.
- Greater provider involvement in planning, quality initiatives, and the work of clinical information coalitions/committees can benefit the organization and provide a different kind of power and satisfaction to clinicians.

5.2 Introduction

Power is "the ability to influence someone else," while *influence* can be defined as "the process of affecting the thoughts, behavior, and feelings of another person."[1] Power is closely related to leadership, authority, hierarchy, and control. There are many theories about sources of power in organizations, but most are based on that described by French and Raven:[2]

- *Reward* power: control of rewards one sees as valuable.
- *Legitimate* power: based on position and mutual agreement.
- *Referent* power: personal power based on liking and respecting someone.
- *Expert* power: personal power based on expertise.
- *Information* power: having access to and control over information (added by Robbins[3]).

Power can certainly be abused. In actuality, however, it is the reason most work gets accomplished—so it also may rightly be viewed positively.[4]

5.3 How Does CPOE Change the Power Structure in Organizations?

In general, we found that all of the sources of power outlined by French and Raven, plus information power, were related to CPOE implementation and use. Legitimate reward (and punishment) power was clearly used by administration when CPOE use was made mandatory. Referent power was exerted by clinical opinion leaders and champions when, because these clinicians were liked and respected, others followed them in using the system. Clinicians themselves hold expert power to such a degree that they can sometimes refuse to use the system or can convince other healthcare workers to enter orders on their behalf. Finally, information power is held by information technology and administrative staff and is evidenced by their having information about clinician ordering patterns or restricting access to data for research purposes.

We found three patterns, all of which were shifts: (1) shifts in the power structure through forced work redistribution and mandated safety pursuits; (2) shifts in control with a perceived loss of clinician control; and (3) shifts in autonomy and a move toward coalitions.

5.3.1 Shifts in the Power Structure

The power held by hospital administrators is formal and carries the ability to influence others, thereby bringing about change. CPOE enables those who hold power to redistribute work and make changes as needed to assure safety.

Forced Work Redistribution

Usually, policy changes—and sometimes even changes to an organization's bylaws—are needed when CPOE is implemented; in turn, they force changes in ways of doing work. Power is often related to the ability to affect the workflow of others. Because administrators generally make the decision to implement CPOE, they are responsible for any shifts in work caused by CPOE. Discussing planned shifts in work responsibilities, one administrator mused:

> Who is going to be the arbiter of that negotiation and who is the power, but also who has the sense of the organization as to what's fair? There is a lot of shifting of work that takes place, and that is a real problem.

One example of an inadvertent shift is that, because clinicians usually enter orders for several patients into the system instead of jotting handwritten orders after each encounter with a patient, they may sit for long periods isolated at a computer:

> You used to have the respiratory therapists round with the doctors and nurses . . . we have a big change now. The doctors have to write the orders into the computer, and the RT isn't usually there, so they are either asking the nurses what the settings are or they have to go find the RT or go check the ventilator themselves.

Mandated Changes for Safety Pursuits

Many unintended consequences related to power shifts are caused by implementation of clinical decision-support systems. CPOE can enforce clinical practice guidelines in many ways, including requiring that particular fields be used for entering data, having default selections in lists that emphasize the least expensive option, and tracking the ordering patterns of individual clinicians to find out if they are practicing according to recommended guidelines:

> There's all these rules that come from pharmacy . . . this has become a political power thing, is how I read it, and they get all emotional, oh, this is for patient safety!

Clinicians are sometimes wary of these measures:

> He used the data for the purpose that he publicly proclaimed he was going to use it for but . . . in the mind of some people these data are being gathered for purposes that people aren't being public about.

Although information technology staff members merely implement major administrative decisions, they are viewed as powerful. A clinician, speaking of safety concerns, said:

> We've said in the past that CPOE should be a clinical project and not an IT project, but it's still amazing how much I think it comes from the IT department.

5.3.2 Perceived Loss of Control and Autonomy

Control is based in power, but is somewhat different in that it also infers monitoring and decision making to alter another person's course of action. Much of the monitoring involves controlling behavior in both overt and subtle ways.

Loss of Control

Alert fatigue is an unintended consequence of decisions that users think are made by administrators (although usually coalitions make the actual decisions). One informant stated:

> An administrator could think that it's good— it's okay to send out 20 alerts if there's one that's going to be right on target—but if you ask the physicians to vote on that and if they're not employed by the institution, they're not going vote [in favor of] a 20-to-1 ratio.

Another way that administration might control users with CPOE is by providing cost information about tests and medications to discourage use of more expensive options. One user noted, however, "I know this lab [test] costs more, so I just ignore the other costs of the labs." In another cost-related discussion, a clinician expressed resentment about selection of the system:

> They went off and got a system with little or no physician input, got something that few docs wanted and ignored the rest of us. This was totally administration-driven to drive down costs.

Control even extends to choice of terminology, which influences reimbursement fees:

> I didn't realize how important nomenclature was in ordering . . . who drives naming them? Is it driven by your interfaces with your other systems? Is it driven by your coders and how you need to charge? Who exactly determines the nomenclature is a big deal.

Nurses have subtle but effective ways of controlling physicians:

> Most nurses are good at encouraging doctors to enter their own orders.

> She will take phone orders from physicians when they are in their offices; if the physicians are on the unit, they are responsible for their orders.

A nurse at a nonteaching hospital said, "The MDs are our guests and can't be ordered to do things." A pharmacist noted:

> "Wrong patient" errors happen daily. Nurses ask physicians to re-enter all of these orders . . . when errors are made by physicians, they are politely pointed out but infrequently corrected on their behalf—this is seen as a learning opportunity.

Because power and control are sometimes informal and based on referent power (you like and respect someone), they can be held by individuals in positions that are not usually considered powerful. For example, ward clerks can informally exert a good deal of control:

> Ms. X [the ward clerk] learned to do it [enter orders]. People are afraid of Ms. X, in a good way. Ms. X runs the unit. You know, she's in control, and so you get somebody like her comfortable with the system and that can sell the merits.

Frontline information technology staff members have subtle strategies for influencing physicians as well. One information technology worker said:

> One of the mistakes we made, we decided to hide their order sheets. We thought, well, we're just going to make this really difficult. If we're not going to mandate it [CPOE], we'll at least hide their order sheets. Well, that was a mistake, because what happened was, the ward clerk got orders on a napkin. You can't have a napkin in the medical–legal chart.

Finally, at sites that had commercial systems, we were told how much control the vendor exerts over the hospital: "We're dependent on the vendor to fix things that are outside the scope of our control."

Loss of Clinician Autonomy

Autonomy implies independence—that one's actions are one's own choices. Physicians have traditionally been highly autonomous because they have been so well respected as the ultimate decision makers in clinical situations. Today their autonomy is being challenged in many ways by the healthcare system as a whole and by the adoption of CPOE in particular. As one interviewee stated explicitly:

> The whole issue of physician autonomy, we don't talk about it much but it needs to be recognized because it's cherished by physicians . . . whether it's the government, whether it's payers, or whether it's CPOE, they resist lessening of their autonomy.

In many respects, CPOE is a threat to the autonomy of the providers who are expected to use the system. Likewise, it poses a threat to some other healthcare professionals, such as nurses and pharmacists. Often, physicians do not directly rebuff administrative power but instead use their power over nurses:

> The physicians are saying, "Well, I didn't go to medical school to be a secretary," and the nurses are saying, "Well, I didn't go to nursing school to be a secretary,"—and the unfortunate thing is the buck always stops with the nurse.

Decision support can be a threat to physician autonomy. We were told a story about another hospital:

> The mindset of the medical staff [was] very much of the belief that they—they—were in control of the hospital, and the practice of medicine in that community . . . so that's why the administration was subversive [about CPOE] because they really believed if they went at it on top of the blanket, it never would have gotten anywhere. The subversiveness was considered necessary, given the culture of that community.

Alert fatigue—a reaction to receiving too many alerts—may partially result from annoyance about being told what to do. One administrator noted:

> They [physicians] view them [alerts] as a nuisance. They don't want to be told how to practice; they don't want a system to suggest practice.

Coalitions Gain Power

While individual physician autonomy is weakening to an extent that surprises clinicians, clinical information systems committees of many kinds are gaining power. Coalitions—groups whose members band together for support so that they can influence other people—have been described as an effective power tactic.[3] The unintended consequence here is that with the proper composition, these coalitions can become extremely powerful. They are often the vehicles for implementing clinical decision support. All of the study sites that had successfully implemented CPOE systems had established interdisciplinary committees to provide oversight of clinical systems, including CPOE. Although these committees include clinicians as members, other clinicians sometimes remain skeptical of their intent, thinking that too much power is given to the committees:

> There are committees to create order sets for each specialty based on best practices and they say, "This is what you'll use," and there's very little way for people to get around that—and I don't like that, I don't trust them.

Conversely, it seems that the clinicians who are most accepting of decisions to implement decision support are closely tied to these committees and highly involved in making decisions about what should be implemented.

5.4 Discussion

Our goal was to gain a deeper understanding of the changes in the power structure that take place when CPOE has been implemented. There are definite shifts in power and control and a loss of individual physician autonomy taking place. The unintended consequence is that the shifts are surprises, especially to the clinicians most affected by them. As the redistribution of power occurs, its ownership becomes less clear, causing discomfort among clinicians and confusion among healthcare team members. Such blurring of role boundaries as a result of information technology has been noted by others. Reddy et al. found that introducing a wireless alert pager system in an ICU changed hierarchical boundaries: It caused shifts in control and blurring of responsibility when an attending physician and a resident received the same alert for the same patient, for example.[5] Saleem et al. warn that role confusion between nurses and providers can become a problem when implementation strategies like clinical reminders are used.[6]

Legitimate hierarchical power is exerted by decision makers at the highest level when they make the decision to implement CPOE. CPOE changes work practices in different ways depending on the system, and those in the organization make many decisions over time that affect workflow. Some decisions are made not *because* of CPOE, but they are *enabled* by CPOE. Users often perceive that IT and quality assurance personnel have gained power—IT because it carries out the enabling process and QA because CPOE enables safety measures that are not possible in a paper-based system. CPOE is the enabler of power redistributions—not necessarily the cause—so blaming CPOE, and especially the technical aspects of the system, is misplacing the culpability. Rather, a better understanding of what is really going on might prove more productive. One strategy for avoiding surprises is to clarify roles and describe changes in power that may occur when CPOE is implemented as part of the planning process.

Whether something is intended or unintended depends on one's perspective. The "unintended consequences" related to the use of power, through CPOE, to make the hospital safer, are unintended from the clinical point of view, but perhaps fully intended by those in power. It is striking that the consequences we found in our study closely mirror the recommendations outlined by Amalberti et al. for lowering barriers to achieving ultrasafe health care: limiting clinician discretion and autonomy; moving toward an equivalent actor mindset (no star surgeons, for example); optimizing safety strategies; and simplifying rules.[7] If these are organizational goals, they need to be understood by all stakeholders. A useful strategy would be to involve clinicians in the process, because only then will they be able to identify what they will gain and lose in the process.

The control mechanisms we saw at our study sites were softer, gentler, power shifts. The doctor–nurse game, which was much discussed in the 1960s,[8] seems to have changed with CPOE. Instead of a strict hierarchy between doctor and nurse, the boundaries are breaking down. In many cases, the nurses become the CPOE experts, but they encourage the physicians to enter their own orders. Nursing expert power is increasing. Mechanic et al. noted control over workplace issues is especially important to physicians, but that control is eroding.[9] Loss of autonomy is one of the main reasons for physician discontent,[8] as CPOE forces guideline and rule adherence. Again, Amalberti et al. have said that other industries have experienced a reduction in autonomy along with improved safety,[7] so they would consider this outcome to be a positive consequence of CPOE. Coalitions of an interdisciplinary nature in the form of clinical information systems committees are powerful and have great influence over CPOE. DiPalma, in an insightful study of power in hospitals, describes these complex coalitions as webs that crisscross hierarchical boundaries, and she calls the result "webbed power."[10] A final strategy would be to encourage such webs.

5.5 Conclusion

Concomitant with CPOE implementation and use is a shift in the power structure of the organization. Much of the redistribution of power is made possible by the ability of the system to provide clinical decision support, which can be used both to monitor and to guide clinician behavior. Because information technology and quality assurance staff tend to gain power as CPOE succeeds, clinicians perceive that they are losing power and autonomy with adoption of such a system. Coalitions are now being formed that have considerable decision-making power over clinical systems and decision support. Clinician involvement in planning for the shifts is needed, and heavy involvement in the work of the coalitions will, in fact, give power to clinicians once again, albeit in a different form. Once the shifts in the power structure become acknowledged and anticipated, they can be managed.

Questions to ask regarding power issues in relation to CPOE include the following:

1. Does the organization have an "information security officer" responsible for implementing and monitoring compliance with all security policies and procedures?

2. Does the organization have a standing committee charged with establishing and revising policies and procedures for protecting patient privacy and for ensuring the security of all information systems?

3. Are all passwords associated with role-based access—that is, narrowly defined roles (e.g., emergency room nurse or cardiologist is preferred over the generic "doctor" or "nurse") for all clinical applications?

4. Does the CIS safety oversight committee keep a record of all errors identified, including how they were identified, how important they are, what will be done to remedy the problem, and who is responsible for making sure the problem is fixed? This record should be reviewed and updated at each meeting.

5. Is there is a single, clearly identified CPOE project leader with realistic attitudes about what can and cannot be accomplished, knowledge of how to educate administrators, and skills to foster teamwork?

6. Is there a clinical information system safety committee (which could be a subset of the CIS oversight committee) that reviews all user reports of system problems and errors resulting from use of the system?

7. Is there executive accountability for major clinical information system projects?

8. Are there regular (monthly) meetings of CIS leadership (e.g., CMIO, CIO, chief nurse) with CEO-level executives?

9. Have the organization's bylaws been reviewed, and modified if necessary, to support the new CIS-driven clinical processes?

5.6 References

1. Nelson DL, Quick JC. *Organizational Behavior: Foundations, Realities, and Challenges.* New York: West Pub; 1994:330.

2. French JRP, Raven B. The bases of social power. In: Cartwright D, Ed. *Group Dynamics: Research and Theory.* Evanston, IL: Row, Peterson; 1962.

3. Robbins SP. *Organizational Behavior.* 11th ed. Upper Saddle River, NJ: Pearson; 2005:392.

4. Block P. *The Empowered Manager: Positive Political Skills at Work.* San Francisco, CA: Jossey-Bass; 1988.

5. Reddy MC, McDonald DW, Pratt W, Shabot MM. Technology, work, and information flows: lessons from the implementation of a wireless alert pager system. *J Biomed Inform.* 2005;38:229–238.

6. Saleem JJ, Patterson ES, Militello L, Render ML, Orshansky G, Asch SM. Exploring barriers and facilitators to the use of computerized clinical reminders. *J Am Med Inform Assoc.* 2005;12:438–447.

7. Amalberti R, Auroy Y, Berwick D, Barach P. Five system barriers to achieving ultrasafe health care. *Ann Intern Med.* 2005;142:756–764.

8. Stein LI. The doctor–nurse game. *Arch Gen Psychiat.* 1967;16:699–703.

9. Mechanic D. Physician discontent: challenges and opportunities. *JAMA.* 2003;290:941–946.

10. DiPalma C. Power at work: navigating hierarchies, teamwork and webs. *J Med Humanities.* 2004;25:291–308.

6 | The Nature of Clinical Information System–Related Errors

Based on: Ash JS, Berg M, Coiera E. Some unintended consequences of information technology in health care: the nature of patient care information system related errors. *J Am Med Inform Assoc.* 2004;11:104–112.

6.1 Key Points

- Clinical information systems (CISs) are complicated technologies, often encompassing millions of lines of code written by many different individuals.

- When such technologies become an integral part of healthcare work practices, we are confronted with a large sociotechnical system in which many behaviors emerge out of the sociotechnical coupling, and the behavior of the overall system in any new situation can never be fully predicted from the individual social or technical components.

- With a heightened awareness of these issues, informaticians can educate others, design systems, and conduct research in such a way that we may be able to avoid the unintended consequences of any subtle, silent medical errors that occur as a result of CPOE systems.

6.2 Introduction

Medical error reduction is an international issue. In 2000, the Institute of Medicine's report on medical errors[1] dramatically called attention to dangers inherent in the U.S. medical care system, which are believed to cause as many as 98,000 deaths in hospitals and cost approximately $38 billion per year. In the United Kingdom, the chief medical officer of the newly established National Patient Safety Agency estimates that "850,000 incidents and errors occur in the NHS [National Health Service] each year."[2] In the Netherlands, the exact implications of the U.S. figures for the Dutch healthcare scene are much debated. Nevertheless, patient safety is on its way to becoming a political priority in that country as well. Medication errors alone have been estimated to cause 80,000 hospital admissions per year in Australia, costing $350 million.[3]

In much of the literature on patient safety, clinical information systems (CISs) are lauded as one of the core building blocks for a safer healthcare system.[4] CISs are broadly defined here as applications that support the healthcare process by allowing healthcare professionals or patients direct access to order-entry systems, medical record systems, radiology information systems, patient information systems, and more. With fully accessible and integrated electronic patient records, and with instant access to up-to-date medical knowledge, faulty decision making owing to a lack of information can be significantly reduced.[5] Likewise, computerized provider order entry (CPOE) systems and automated reminder systems can reduce errors by eliminating illegible orders, improving communication, improving the tracking of orders, checking for inappropriate orders, and reminding professionals of actions to be undertaken. In this way, these systems can contribute to preventing underuse, overuse, or misuse of diagnostic or therapeutic interventions.[6–8] Among the broad array of health informatics applications, CPOE systems—and especially medication systems—have received the most attention to date.[9–12]

CISs are complicated technologies, often encompassing millions of lines of code that are written by many different individuals. The interaction space[13] within which clinicians carry out their work can also be immensely complex, as individuals can execute their tasks by communicating across rich social networks. When such technologies become an integral part of healthcare work practices, we are confronted with a large sociotechnical system in which many behaviors may emerge from the sociotechnical coupling, and the behavior of the overall system in any new situation can never be fully predicted from its individual social or technical components.[1,13–17]

It is not surprising, therefore, that authors have started to describe some of the unintended consequences that the implementation of CISs may trigger.[18] For instance, professionals may trust the decision support suggested by the seemingly objective computer more than is actually called for.[15,19] CISs may impose additional work tasks on already heavily burdened professionals,[20,21] and the tasks are often clerical in nature and, therefore, economically inefficient.[17] These systems may also upset smooth working relations and communication routines.[13,22] Likewise, given their complexity, CISs may themselves contain design flaws "that generate specific hazards and require vigilance to detect"[23,24] As a consequence, CISs may not be as successful in preventing errors as is generally hoped. Even worse, CISs may sometimes actually *generate* new errors.[25–27]

It is obvious that CISs will ultimately be a necessary component of any high-quality healthcare delivery system. Yet, in our research in three different countries, we have each encountered many instances where CIS applications seemed to *encourage* errors rather than *reduce* their likelihood. In healthcare practices in the United States, Europe, and Australia alike, we have seen situations in which the system of people, technologies, organizational routines, and regulations that constitutes any healthcare practice seemed to be *weakened* rather than *strengthened* by the introduction of the CIS application. In other words, we frequently observed instances where the intended strengthening of one link in the chain of care actually led unwittingly to a deletion or weakening of other links.

In this chapter, we argue that many of these errors are due to highly specific failures in CIS design and/or implementation. We will not focus here on those errors that are attributable to faulty programming or other technical dysfunctions. Hardware problems and software bugs are more common than they should be—especially in a high-risk field such as medicine. Nevertheless, these problems are well-known, and can theoretically be eliminated through testing before implementation. Similarly, we will not discuss errors that are attributable to obvious individual or organizational dysfunction, such as a physician refusing to seek information in the computer system "because that is not his task" or a healthcare delivery organization cutting training programs for a new CIS for budgetary reasons.

Instead, in this chapter we focus on those often latent or silent errors that are caused by a mismatch between the functioning of the CIS and the real-life demands of healthcare work. Such errors are not easily found by a technical analysis of the CIS design, nor may they even be suspected after the first encounter with the system in use. Rather, these errors emerge only when the technical system is embedded into a working organization, and may vary from one organization to the next. Yet in failing to take seriously some by now well-recognized features of healthcare work, *some CISs are designed or implemented in such a way that errors can arguably be expected to result.*[28] Only when thoughtful consideration is given to these issues, we argue, will CISs be able to fulfill their promise.

6.3 Background and Methods

Information contained in this chapter draws upon a literature review and a series of qualitative research studies conducted in the United States, the Netherlands, and Australia. These studies utilized standard qualitative methods such as ethnographic observation in healthcare delivery settings and semi-structured interviews with professionals.[29] All of the studies focused on the effects of CISs on health care, yet none of the researchers deliberately set out to study error prevention. We did not focus especially on problematic CIS implementations; on the contrary, most of our studies were performed at sites that were recognized as highly successful healthcare facilities. While discussing our different research projects, however, we realized that we had all gathered data that indicated the possibility of, fear about, or awareness of CIS-related errors. In our view, the importance of this topic, and the relevance of the lessons these data can teach us, warrant a blended international treatment of this issue.

Because these investigations were qualitative studies, they did not offer estimates of how *often* certain errors occurred, nor did they indicate whether CISs overall result in more or fewer medication errors, for example. Rather, the power of qualitative work lies in the richness of its detailed descriptions.[29] We include results from the underlying studies, from diverse fields in diverse contexts, to emphasize the ubiquity of the issues addressed in this chapter. We offer an interpretation of the *nature* of healthcare work, the role of information and information technology, and the risks of an improper interrelation or fit between CISs and healthcare work. Our goal is to present an argument, supported by extensive literature and illustrated by prototypical examples from our studies, that will lead to more quantitative work that can, in turn, track the "epidemiology" of these information system pathologies, as well as convince decision makers to be cautiously realistic about the benefits of CISs.

The following discussion includes verbatim quotes from interviewees or field notes that illustrate patterns seen across the studies. The quotes were selected because they are both representative and well-stated. It is important to emphasize that, as far as we are aware, the examples detailed in this chapter never led to actual harm of patients.

6.4 Kinds of Silent Errors

The complex nature of healthcare work both creates and hides errors, which can be nearly invisible or silent. Healthcare work can be characterized as the managing of patients' trajectories: Under continuous time pressure, and in constant interaction with colleagues and the patient, healthcare professionals try to keep a patient's problem on track. This understanding implies that healthcare professionals simultaneously act on a whole range of dimensions, including interpreting physical signs and diagnostic tests, and dealing with organizational policies and the patient's individual needs. However standardized the diagnostic and/or therapeutic care paths, individual patient trajectories always follow their own, unique course. Contingencies are the rule; smoothly molding such continuous lapses of order into events to be handled with "standard operating procedures" is the true skill of experienced healthcare professionals.[13,30,31] Computer applications are best when they automate routine work, but the complexities of the healthcare process often make it anything but routine.

In outpatient settings, the interaction with colleagues may be less pronounced than in inpatient settings, although professionals in inpatient settings are certainly connected through other healthcare professionals' opinions and needs through progress reports and referrals. This social organization of medical work, as the sociologist Strauss called it, is now widely recognized as an important feature to consider in the design of healthcare information technologies.[31–35]

In this section, we discuss two main categories of errors that occur at the interface of the information system and work practice that are attributable to a failure to grasp this nature of healthcare work. First, we discuss errors in the process of *entering and retrieving* information in or from the system. Second, we discuss errors in the *communication and coordination* processes that the CIS is supposed to support. As the examples presented here illustrate, such failures are caused by mistaken assumptions about healthcare work that are built into CIS applications, which create dysfunctional interactions with users and, sometimes, lead to actual errors in the delivery of healthcare services.

6.4.1 Errors in the Process of Entering and Retrieving Information

Increasingly, the entry and retrieval of information into and from a CIS is a core activity in healthcare delivery. Given the characteristics of this work, these CIS applications have to fulfill specific demands. Many of these criteria are well-known: CIS applications have to have fast response times, have negligible downtime, be easily accessible, and be accessed through interfaces that are easy to understand and navigate.[5,36] Also, the software and hardware must be designed to optimally fit the ecology of the work practice: mobile when necessary, robust and small, yet ergonomically suitable.[13,37] Although such requirements are widely known and accepted, they are often not met. Many system interfaces are still so impractical that using the systems takes a great deal of (costly) time on the part of busy professionals. Some systems in use in medical practices today have interfaces that are outdated, with no windows, no intuitive graphical navigation aids, and endless lines of identical-looking text. In such cases, even when the information is there, it may be exceedingly difficult to find. Here, we discuss two problems in detail:

- CISs that have human–computer interfaces that are not suitable for this highly interruptive use-context
- CISs that cause cognitive overload by over-emphasizing structured and complete information entry or retrieval

Human–Computer Interface That Is Unsuitable for a Highly Interruptive Use-Context

Working on the computer is rarely an isolated task: Healthcare professionals are always communicating with others, including patients in outpatient settings, but primarily with other healthcare professionals. More often than not, different tasks are executed simultaneously, and interruptions by beepers, telephones, and colleagues are endless.[38,39] Many human–computer interfaces, however, seem to have been designed for workers doing their work by themselves, fully and extensively concentrating on the computer screens. This single-task assumption is aggravated by the fact that so many existing screen designs are already suboptimal by current office standards.

This mismatch between interface and use-context often results in a *juxtaposition* error: the kind of error that can result when something is close to something else on the screen and the wrong option is too easily clicked in error. The following are typical quotations from physicians (note the allusions to the "interruptive" use-context):

I have ordered the test that was right next to the one I thought I ordered—you know, right below it that my little thingie [the pointer controlled by the mouse] had come down and I clicked and I'm looking at this one, but I in fact clicked on the thing before. By that time I turned my head and I'm hitting Return and typing my signature and not seeing it. (Physician, U.S. hospital)

I was ordering Cortisporin, and Cortisporin solution and suspension come up. The patient was talking to me; I accidentally put down solution and realized that's not what I wanted . . . I would not have made that mistake, or potential mistake, if I had been writing it out because I would have put down what I wanted. (Physician, U.S. outpatient setting)

Likewise, there were many instances of patient or physician confusion when *orders were entered for or on behalf of the wrong person.* Again, in a context of many co-occurring activities and interruptions, a suboptimal interface becomes rapidly unforgiving:

Patients were getting the wrong orders for medications. You would order it on one patient and it [was], because of the vagaries of the light-pen system, . . . really ordered on somebody else—and somebody got the wrong medication and that sort of thing. (Physician, U.S. hospital)

She looked up the patient's diet and was trying to order a regular diet. At the fifth screen she saw that the patient was getting tube feeding. This clued her that this was the wrong patient. (Field notes, observation of a nurse, U.S. hospital)

Cognitive Overload Caused by Over-emphasizing Structured and "Complete" Information Entry or Retrieval

Professionals need fast access to data that are relevant to the case at hand. Simultaneously, they need to be able to record a maximum amount of information in a minimum amount of time—and in such a way that it is most useful to other healthcare professionals involved in the handling of this patient's trajectory. Psychological and sociological studies have shown that in a shared context, use of concise, unconstrained free-text communication is most effective for coordinating work around a complex task.[40–42] Attempting to require professionals to encode data, or to enter data in more structured formats, can be fruitful, and is necessary for research or managerial purposes, but comes with a real cost: Such formats are generally more time-consuming to complete and read. When the information's relevancy to the primary task is lessened through the structuring of the information, and/or when the time spent writing or reading this information increases significantly, the information ends up being *less* useful for the primary task at hand.[20]

Structure.   Some CIS systems require data entry that is so elaborate that the time spent recording patient data is significantly greater than it was with these systems' paper-based predecessors. What is worse, on several occasions during our studies, overly structured data entry led to a loss of cognitive focus by the clinician. Having to go to many different fields, often using many different screens to enter many details, physicians reported a *loss of overview*. When professionals are working through a case—determining a differential diagnosis, for example—the act of *writing* the information is integral to the cognitive processing of the case.[43,44] This act of writing-as-thinking can be aided greatly by *some* structure, such as the grouping of similar types of information or sequencing to guide elucidating a history, but it is inevitably hampered by an excess of structure. Rather than helping the physician build a cognitive pattern to understand the complexities of the case, such systems overload the user with details at odds with the cognitive model the user is trying to develop.

Fragmentation.   Similarly, the need to switch between different screens can result in a loss of overview. Physicians and nurses in an intensive care unit, for example, reported that the large paper day-sheets they used to work with would include an order list, problem list, vital signs graphs, and medication lists—all on a single large sheet of paper. The graphic user interface software they used supported all of these functions and more, but the user had to switch among multiple windows to obtain all of the information. Doing so, several professionals argued, worked against their ability to acquire, maintain, and refine a mental overview of the case. Some reported that they felt insecure about identifying emerging problems because the activity of clicking through the different screens inevitably fragmented the cognitive "images" they were constructing.

Likewise, records might overly separate the information flows according to work task or responsibility. In everyday practice, doctors may gather information from nurses' notes, or the notes of other specialists, that relate to the problem. Information systems may limit this easy access to other people's notes or other parts of the record, thereby severely hampering the professional's ability to be optimally informed.[45]

> [R]egarding interpretation of results, currently this is often in the notes so [you] can see the results and the interpretation. On an order entry, results reporting may only get the raw data and not the interpretation, which could affect clinical work. This separation may also lead to clinicians being too specialty focused [and] not seeing what others have written—now [we] have to flick through notes so we see other information. On this system, if [we] only go to [our] own information, this may not happen and information may be missed. (Allied health professional, Australia)

Over-completeness. Results reporting systems may also mistake completeness for efficacy. In several instances, physicians stated that systems that produced standard, "complete" reports actually reduced the usability and the transparency of these reports or discharge letters. The physicians explained:

> There are so many standard phrases in the ordinary reports; I don't think that's good ... you have to really search for the usable information ... Many others use the [standard templates] and then you often see a discussion with standard phrases, one or two added phrases, and then more standard phrases. You then have to really search what the considerations were ... In my reports the text is mine; it doesn't come from the computer—I make it up myself ... Everyone should do that. If you have so much standard text, it becomes too easy to just push that button and add some more. (Insurance physician, the Netherlands)

> You'll have to write the largest part yourself. You can standardize only so much, since otherwise you get an empty report with only standard phrases that could be true for anyone. (Insurance physician, the Netherlands)

Too many standard phrases, these physicians argued, actually decreased the readability and information value of the reports. From the point of view of the professional, overly "complete" reports may end up becoming "empty" and stand in the way of actual communication. The similarity of the phrases, and the impossibility of judging whether a sentence is part of the template or the result of a thoughtful weighing of words, threatens to obscure the transparency that such systems attempt to introduce.

Of course, ease of use can also lure users into learning new but poor recording practices. The ability to cut-and-paste data or, more often, copy-and-paste data, affords users the opportunity to exacerbate the data overload problem. As an attending physician stated:

> Just before I came up here, I looked at a discharge summary that was an absolute disaster, because not only had she cut and pasted the progress note, but she had cut and pasted the whole thing, so the intern's signature and the whole thing were on it. [The system] is inherently error prone ... people have the tendency to cut and paste ... and instead

of taking the pertinent facts from a laboratory report or from another clinician's progress note, they will cut and paste a whole laboratory report, cut and paste somebody else's thinking process into their own note, and sign it. (Physician, U.S. hospital)

6.4.2 Errors in the Communication and Coordination Process

The previous section discussed errors related to the processes of entering and retrieving information in CISs. In this section, we focus on the way computers can undermine communication about and coordination of events and activities. Here we encounter the truly interactive and contingent nature of healthcare work and the consequences of not taking these characteristics into account. Although the issues to be discussed here are highly interrelated, we have subdivided them into two overarching problems:

- Misrepresenting collective, interactive work as a linear, clear-cut, and predictable workflow

- Misrepresenting communication as information transfer

Misrepresenting Collective, Interactive Work as a Linear, Clear-Cut, and Predictable Workflow

CIS systems often appear to be imbued with a formal, stepwise notion of healthcare work: A doctor orders an intervention, a nurse subsequently arranges for or carries out the intervention, and then the doctor obtains the information about the result. As a chain of independent actions, an order is executed and reported upon, or a piece of information is generated, processed, and stored.[46–48] Yet it is common knowledge that it is inherently difficult for formal systems to accurately handle or anticipate the highly flexible and fluid ways in which professional work is executed in real life.[13,49,50] Care path or workflow systems are plagued by the ubiquity of exceptions.[51] Similarly, decision-support systems are in constant need of "supervision" to determine whether their suggestions fit a given case.[18] Systems cannot handle all potential exceptions: Very soon, the number of branching points becomes too great, and the system becomes impossible to maintain and to use.[52]

Support of work processes is one of the main benefits of CISs, yet brings its own problems. Finding the proper balance between formalizing work activities so that the information technology application can fulfill its promise, and respecting this fluid and contingency-driven nature of healthcare work, is no easy task for system designers.[53] Nevertheless, it is a necessary endeavor if CIS systems are to contribute to the overall quality improvement required in Western health care.

Inflexibility.   Current CISs often fail to reflect some of the basic real-life exigencies of care work, thereby resulting in problems for the user, and potentially faulty reporting and actions. Seemingly easy and clear-cut on paper, the real-time intricacies of treatment protocols, for example, may baffle the system's preconceptions of these processes. In one instance, for example, a drug ordered three times a day (isordil) had been discontinued, but one dose had already been given. The computer system would not allow the nurse to chart the one dose, because the system considered it an incomplete execution of the task. (Pharmacist at a U.S. hospital and recorded in field notes)

Urgency.   In the case of urgent medication orders, nurses may give a medication before the physician formally activates the order. A familiar category of errors arises in this situation that has to do with the informal realities of medication handling in health care. In everyday healthcare work, experienced nurses often have more practical knowledge about which medications to give when, and which contraindications may be relevant, than many of the junior physicians who populate the wards.[54] For example, during nightly routine medication administration, nurses may initiate distribution of the medications without waking up the junior doctor who is formally responsible for signing the order. There is a rather large gray zone of informal management of these responsibilities and tasks, which can be entirely rational, given the everyday organization and exigencies of healthcare work. Within this same gray zone, there may lie many practices that would contribute to unsafe medication routines, such as doctors actively discouraging nurses from calling them for medication requests, or nurses taking too many liberties with dosing.

All of these practices exist within the current paper-based medication systems, but many computerized medication systems all too radically *cut off* such practices. As a consequence, many computerized medication systems have been rejected by their users because they strictly demanded a physician's authorization *before* any drug could be distributed, or because they made any alternative route (such as the nurse ordering the medication through an "agent-for" procedure) much too cumbersome. In the last example, nurses had to bear the consequences of physicians not wanting to have to enter every medication order before anything could be given or changed. Understandably, both professional groups refused to fulfill these demands.[55]

Workarounds.   When such systems do remain in practice, workarounds, which are clever alternative approaches, are artfully developed by the users. Workarounds allow users to live with the system while avoiding some of the demands that are deemed unrealistic or harmful.[56,57] Such situations may undermine patient safety, however. In urgent situations, physicians may enter medication orders *after* the medication has already been administered, for example. Alternatively, the order might have been entered by the nurse but would have to be activated by the physician on a post hoc basis. One critical care nurse in a U.S. hospital remarked that in such situations, near the change of her shift, she often "worries that the [urgently given] medication may be given again when the order is activated."

Transfers.   Similar problems abound when transferring patients between wards, or when admitting new patients. Here again, the real patient flow does not always match the clear-cut, formal model of the patient flow, which begins with the input of the required administrative data, after which the clinical content can be accessed and entered. This model is intended to ensure that the patient record is not accidentally fragmented over different electronic patient-identities. In real-life health work, however, information may be required or activities may be started or planned *before* the proper administrative details are entered or even known. Such problems are familiar to everyone with some clinical experience, yet some CPOE systems very poorly support this practice, as we witnessed in all three countries in our study. For example, during transfers between the emergency department and a patient ward, orders would not be transferred, or new orders could not be entered in the system, because the patient was not yet "in the system." "If they don't remember or know their Social Security number, it's tough," a U.S hospital nurse remarked.

In another example, we were told that once an order had been entered by a physician, that person expected it to be carried out. If the administrative data had not yet been entered, however, the physician's orders might never be executed. "The doctors liked to be able to write orders and hold them pending an admission and the software was dropping off the orders you know . . . that was just incredible." (Nurse, U.S. hospital)

A similar issue is "the midnight problem." It does not make a large difference for ongoing practical work or for a patient's health whether an order is issued just before or just after midnight, but some systems create a difference. This may make sense from a purely administrative perspective but not from a clinical one. "If the patient has a while to wait in the ER [Wednesday night] for a bed, or some other delay and doesn't get on the floor until 12:01 A.M. [Thursday], the order [for tomorrow's medications] effectively means Friday morning. Big problem from his perspective, and I heard this from two other doctors as well. This would cause there to be no orders in the system for Thursday." (Field notes, observation of U.S. physician)

Misrepresenting Communication as Information Transfer

Loss of Communication.   In a work practice such as health care, which is characterized by contingencies and constantly developing definitions of the situation, proper communication among the involved professionals is crucial. However, "physicians may assume that 'entry' into the computer system replaces their previous means of initiating and communicating their plans, and that orders will be carried out without further action on their part. The result is reduced direct interaction among physicians, nurses, and pharmacy and increased overall reliance on the computer system."[58] The entry of information into the system, in other words, is not the same as completing a successful *communication act*.

When a U.K. hospital replaced the practice of laboratory staff telephoning results with installation of a results-reporting system in an emergency department and on the medical admissions ward, the results were devastating: "The results from 1443/3228 (45%) of urgent requests from accident and emergency and 529/1836 (29%) from the admissions ward were never accessed via the ward terminal . . . In up to 43/1443 (3%) of the accident and emergency test results that were never looked at, the findings might have led to an immediate change in patient management."[59] In this case, the designers had overlooked the fact that in the previous work process, laboratory personnel called doctors when the results were in. In the new situation, doctors would have to actively log into the system to see whether the results were available. In the hectic environment of these wards, this proved to be a highly inefficient mode of communication for these professionals.[13]

Loss of Feedback. We encountered many variations on this theme—nurses are often alerted to new orders by the printer, but this assumes the nurse is nearby and that the printer functions correctly:

> There is a printer problem—for example, you know, something prints out or that piece of paper that gets printed out at the nurses' station somehow gets lost or not seen. I've seen a couple of antibiotics get missed. (Physician, U.S. hospital)

Likewise, a typical complaint is that, "He was totally unaware of this new order—he had heard no mention of it previously, and there had not been a notification of the order by the ordering physician." (Field notes, observing a nurse at a U.S. hospital) Here again, the sender of the information mistakenly assumes that the computer will take care of notifying the receiver, the nurse.

Similarly, a common problem is that physicians cannot tell whether an order has been carried out, or whether someone else has entered a similar order, without gaining feedback. In one U.S. hospital, we discovered that nurses put their initials into the computer when they take the order off rather than, for example, when they have completed the order. The latter procedure might be more correct, but it would require yet another separate computer session. Although logical from the nurses' point of view, the system did not make a distinction between an order that was accepted and an order that was executed. This was problematic because doctors then often do not know the true status of orders. (Field notes, observing nurses and physicians at a U.S. hospital)

Due to miscommunication, orders or appointments may be missed, diagnostic tests may be delayed, and medication may not be given. Communication involves more than simply transferring information. Communication is about generating an *effect*—the laboratory personnel want to make sure that the doctors will *act* upon their data. Similarly, communication is about testing out assumptions about the other person's understanding of the situation, and willingness to act upon your information.[60–62] In addition, communication is always about establishing, testing, or maintaining relationships.[63]

Decision-Support Overload. Decision-support systems tend to suffer from the same problem: They may trigger an overdose of reminders, alerts, or warning messages. These messages may be sent to the computer user even if the message is not relevant for that user at that moment or even if the *intended* recipient of the message is not the person who is entering the data. From a communication perspective, it is crucial to realize that such messages may generate a problem much larger than simple data overload[18]; that is, the user may feel supervised, treated as "stupid," distrusted, or resentful of being constantly interrupted.

To deal with this issue, healthcare professionals may disregard the multitude of messages, click them away, or turn the warning systems off whenever they have an opportunity to do so. It is common to blame these professionals for such seemingly irresponsible behavior. However, in too many systems, too little attention is paid to ensuring the judicious use of alerts and to working on the problem of *contextual relevancy* for the alerts the system generates during actual use. When time is a scarce resource, and too many of the warnings or reminders are either irrelevant or overly predictable, irritated physicians who disregard these alerts are acting quite rationally.

Catching Errors. Appropriate and well-supported communication is also part and parcel of a safe work practice. In this sense, the systems we describe in this subsection may actually *hamper* safer working practices rather than stimulate them. In the hierarchies and task divisions of manual ordering, for example, many error prevention mechanisms are built into the ordering process, albeit often informally. For example, pharmacists routinely correct the medication orders given by physicians. Restructuring the medication ordering process might unwittingly eliminate these important mechanisms. "POE [physician order entry] systems founded on notions of individual cognition are likely to be constrained by this model and be unable to take advantage of the distributed processing, fault tolerance, and resilience that obtains in settings characterized by distributed cooperative problem solving."[58] Errors are caught constantly—and not necessarily by those formally responsible for doing so.[64]

The redundancy that is built into the system of people and technologies constituting the medication management chain is partly responsible for the fact that of the many *prescription* mistakes made, only a minute fraction results in actual medication *administration* mistakes. Similarly, in practice, orders often come into being during patient rounds, during discussions among senior and junior physicians and nurses. A case is discussed, a suggestion is made and elaborated on, and it becomes an order. It may also be transformed, renegotiated, or ignored. When details remain unclear, those involved can ask for elaboration, or smoothly "repair" interpretations of junior members of the team. In most clinical order-entry systems, however, the entering of orders is the task of the junior resident who does ordering only *after* the patient rounds. This delay occurs because systems are rarely mobile, so they are not available during rounds. Alone at a computer, the resident enters a series of orders on a series of patients, copying from the notes made during rounds. In such a setting, outside of the actual context in which the patient was discussed, and away from those who may correct the resident's misinterpretations, order entry may be prone to errors.

6.5 Discussion and Conclusion

We have outlined a number of issues within a framework describing two major kinds of silent errors caused by healthcare information systems: those related to entering and retrieving information and those related to communication and coordination. Because the potential causes of these errors are subtle but insidious, the problems need to be addressed in a variety of ways through improvements in education, systems design, and research.

6.5.1 Education

Health professionals need to be educated with a *critical* perspective toward what CISs can do for them. People tend to project "intelligence" and "objectivity" on to computers,[15,65,66] and physicians and nurses are no exception. In the classic case of the Therac-25 system, a computer-controlled radiation machine that was the cause of radiation overdoses in six patients, the operators trusted the "All is normal" messages the machine was delivering. Ultimately, they disregarded disturbing clinical signs because they had faith in the machine.[67] In a study of computer decision support in health care, users were unduly influenced by incorrect advice.[68]

Informatics education has a role to play in preventing these kinds of errors by making sure that clinical systems are designed, implemented, and evaluated with this notion in mind. Medical education—and indeed the education of all healthcare professionals—should involve consideration of both the positives and negatives of using information systems. The outcome of these educational efforts should be a workforce whose members practice appropriate diligence when using a CIS.

In addition, it is imperative that we educate an increasing number of people who can bridge the gap between the clinical and technological worlds, who can speak the language of both and, therefore, can act as translators. These individuals, whom we will call clinical informaticians, need to be included in systems design efforts and in systems implementation. During the implementation process in particular, they need to assure that clinicians are not only heavily involved so that the implementation goes more smoothly, but also that clinicians are able to continue the social processes that the system may supplant. For example, luncheon meetings for the purpose of discussing new functions of the system might replace some of the communication loss caused by a CPOE system.

6.5.2 Systems Design

Systems developers and vendors should be clearer about the limitations of their technologies. When speaking of "order entry" and "intelligent" systems, and building upon overly rationalist models of healthcare work, they can too readily lure users into expecting much more from a computer system than it can actually deliver. Systems should be designed to support *communication*[13] and should provide the flexibility that is needed for systems to better fit real work practices. Many lessons can be learned from proponents of good systems design, and though the technology is rapidly improving, known design principles are still not evident in today's systems. Increasing involvement of experienced clinicians who know what the work is truly like should improve designs of CPOE systems in the future. The hiring of increasing numbers of clinical informaticians by vendors and healthcare organizations alike to design and customize systems is a positive trend. In addition, even systems designers with no clinical experience should seek to spend some time simply observing clinical activities so that the nature of these activities can be experienced first-hand.

Systems should be able to help clinicians manage interruptions, perhaps by reminding them about what they were doing when last using the system. The systems also need more effective feedback mechanisms so clinicians will know if and when the orders are being received and carried out. Mobile systems hold promise for assisting with overcoming these problems related to both interruptions and lack of feedback, so further development efforts should focus on them.

Prevention of silent errors is preferable to fixing such errors after the fact. Repairing these errors by adding safety features that are not thoroughly designed may very well make things worse. Introducing safety devices is an artful process in its own right, requiring thorough insight into the communication space. For example, an observer wrote in his field notes:

> We were told that the answer to this problem was that they inserted a safety level which is yet another screen [all laugh] so that when you press on the patient, then there's five lines of information about this person and you have to verify each one . . . at what point are safety levels [more screens to make sure it's the right patient] more disruptive than helpful—similar problem to having too many alerts or too much information to take in. (Field notes, observation of house officers, U.S. hospital)

Systems designers are not to be blamed for silent errors. Sometimes a problem really could have been anticipated, but others problems are so subtle that they can be identified only by closely monitoring practice. Clearly, constant vigilance is crucial. Information systems, by themselves, are not a sufficient fix for the safety problem. A rush toward implementing systems might ultimately endanger the quality of care more than help it.

6.5.3 Research

In practice, the flow of healthcare work activities is often much less linear than it is in other arenas, with healthcare roles being much more flexibly defined and overlapping, and distinctions between steps being much more fuzzy than the formalized CIS models would have it.[18,69,70] Because of this complexity, standard quantitative research methods such as surveys often fail to expose the subtle problems. Qualitative research techniques, by comparison, can provide deep insight and can both identify problems and answer the "why" and "how" questions that quantitative studies cannot answer.[71] This research needs to be multidisciplinary and consider the multiple perspectives of all stakeholder groups.

Finally, all of us involved in information technology in health care need to practice heightened vigilance. We must be aware of the issues described in this chapter through education and training, be alert to the problems identified through further research, be cautious when making major changes that might have unintended consequences, and be prepared to deal with the inevitability of such consequences. We should also be optimistic: If we can identify the presence of unintended negative consequences early enough, we can do something about them. If we can reach a high enough level of vigilance, we may be able to completely avoid many of the subtle silent errors described here.

Questions to ask regarding new kinds of errors in relation to CPOE include the following:

1. What is the total number of database transaction errors that are occurring on a daily basis?

2. Are there any "test" patients or "test" clinicians in the production or "live" system with the name of "Test," silly names (e.g., Donald Duck, Ima Sick, Pickup Andropoff), or permuted names of famous people (e.g., Jorge Wershington, Albert Ainstein)? Test patients should be clearly named as such. For example, Kaiser Permanente Northwest uses "KPNW23, KPNW23" as the text name, while Brigham and Women's Hospital uses "BWH03, BWH03" for this purpose.

3. Is the test version of the application clearly marked as such (e.g., different background color of application or "TEST" appearing as a watermark on all screens) to prevent clinicians from inadvertently entering data describing real patients that will not be acted upon?

4. Are no more than two instances (or sessions with a different patient) of the clinical information system application allowed to run under a single login on a single workstation? Such a policy should limit the possibility of inadvertent "wrong patient" errors.

5. If a second session is created, is it clearly marked by something as obvious as a different background for the entire application?

6. What percentage of orders (all types: medication, laboratory, and radiology) have an indication associated with the order? Not only is this information of value in the billing process, but it should also help reduce the possibility of "wrong patient" errors.

7. What percentage of free-text data entries are the same length as the maximum buffer length on the field? This measure should be monitored.

8. What percentage of all orders are entered as "miscellaneous" or all free text? This measure should be monitored and reviewed periodically (e.g., quarterly, semi-annually, annually).

6.6 References

1. Institute of Medicine, Committee on Quality of Health Care in America. *To Err is Human: Building a Safer Health System.* Washington, DC: National Academy Press; 2000.

2. Donaldson L, et al. An organisation with a memory: report of an expert group on learning from adverse events in the NHS. Department of Health, United Kingdom, 2000 (108 pgs). Available at: http://www.dh.gov.uk/prod_consum_dh/ groups/dh_digitalassets/@dh/@en/documents/digitalasset/dh_4065086.pdf. Accessed August 16, 2009.

3. Roughead E. The nature and extent of drug-related hospitalizations in Australia. *J Qual Clin Prac.* 1999;19:19–22.

4. Institute of Medicine, Committee on Quality of Health Care in America. *Crossing the Quality Chasm: A New Health System for the 21st Century.* Washington, DC: National Academy Press; 2001.

5. Dick RS, Steen EB, Detmer DE, Eds. *The Computer-Based Patient Record: An Essential Technology for Health Care.* Washington, DC: National Academy Press; 1997.

6. Bates DW, Pappius E, Kuperman GJ, et al. Using information systems to measure and improve quality. *Int J Med Inform.* 1999;53:115–124.

7. McDonald CJ, Hui SL, Smith DM, et al. Reminders to physicians from an introspective computer medical record: a two-year randomized trial. *Ann Intern Med.* 1984;100:130–138.

8. van Wijk M, Van der Lei J, Mosseveld M, Bohnen A, van Bemmel JH. Assessment of decision support for blood test ordering in primary care: a randomized trial. *Ann Intern Med.* 2001;134:274–281.

9. Sittig DF, Stead WW. Computer-based physician order entry: the state of the art. *J Am Med Inform Assoc.* 1994;1:108–123.

10. Bates DW, Leape LL, Cullen DJ, et al. Effect of computerized physician order entry and a team intervention on prevention of serious medication errors. *JAMA.* 1998;280:1216–1311.

11. Dexter PR, Perkins S, Overhage JM, et al. A computerized reminder system to increase the use of preventive care for hospitalized patients. *N Engl J Med.* 2001;345:965–970.

12. Teich JM, Merchia PR, Schmiz JL, et al. Effects of computerized physician order entry on prescribing practices. *Arch Intern Med.* 2000;160:2741–2747.

13. Coiera E. When conversation is better than computation. *J Am Med Inform Assoc.* 2000;7:277–286.

14. Perrow C. *Normal Accidents: Living with High Risk Technologies.* New York: Basic Books; 1984.

15. Weizenbaum J. *Computer Power and Human Reason: From Judgment to Calculation.* San Francisco: W.H. Freeman; 1976.

16. Berg M. Implementing information systems in health care organizations: myths and challenges. *Int J Med Inform.* 2001;64:143–156.

17. Tenner E. *Why Things Bite Back: Technology and the Revenge of Unintended Consequences.* New York: Vintage Books; 1996.

18. Goldstein MK, Hoffman BB, Coleman RW, et al. Patient safety in guideline-based decision support for hypertension management: ATHENA DSS. *Proc AMIA Symp.* 2001:214–218.

19. Burnum JF. The misinformation era: the fall of the medical record. *Ann Intern Med.* 1989;110:482–484.

20. Berg M, Goorman E. The contextual nature of medical information. *Int J Med Inform.* 1999;56:51–60.

21. Massaro TA. Introducing physician order entry at a major academic medical center: I. Impact on organizational culture and behavior. *Acad Med.* 1993;68:20–25.

22. Dykstra R. Computerized physician order entry and communication: reciprocal impacts. *J Am Med Inform Assoc Symp Supp.* 2002:230–234.

23. Shojania KG, Duncan BW, McDonald KM, Wachter RM. Safe but sound: patient safety meets evidence-based medicine. *JAMA.* 2002;88:508–513.

24. Effken JA, Carty B. The era of patient safety: implications for nursing informatics curricula. *J Am Med Inform Assoc.* 2002;9(suppl):S120–S123.

25. Weiner M, Gress T, Thiemann DR, et al. Contrasting views of physicians and nurses about an inpatient computer-based provider order-entry system. *J Am Med Inform Assoc.* 1999;6:234–244.

26. Bates DW, Cohen M, Leape LL, et al. Reducing the frequency of errors in medicine using information technology. *J Am Med Inform Assoc.* 2001;8:299–308.

27. McNutt RA, Abrams R, Arons DC. Patient safety efforts should focus on medical errors. *JAMA.* 2002;287:1997–2001.

28. Han YY, Carcillo JA, Venkataraman ST, et al. Unexpected increased mortality after implementation of a commercially sold computerized physician order-entry system. *Pediatrics.* 2005;116(6):1506–1512.

29. Strauss AL. *Qualitative Analysis for Social Scientists.* Cambridge, UK: Cambridge University Press; 1987.

30. Berg M. Medical work and the computer-based patient record: a sociological perspective. *Meth Inform Med.* 1998;37:294–301.

31. Strauss A, Fagerhaugh S, Suczek B, Wieder C. *Social Organization of Medical Work.* Chicago: University of Chicago Press; 1985.

32. Drazen E L, Metzger JB, Ritter JL, Schneider MK. *Patient Care Information Systems: Successful Design and Implementation.* New York: Springer; 1995.

33. Kaplan B. Objectification and negotiation in interpreting clinical images: implications for computer-based patient records. *Art Intell Med.* 1995;7:439–454.

34. Kuhn KA, Guise DA. From hospital information systems to health information systems: problems, challenges, perspectives. *Meth Inform Med.* 2001;40:275–287.

35. Forsythe DE. *Studying Those Who Study Us: An Anthropologist in the World of Artificial Intelligence.* Stanford, CA: Stanford University Press; 2001.

36. Ash JS, Gorman PN, Lavelle M, Payne TH, Massaro TA, Frantz GL, Lyman JA. A cross-site qualitative study of physician order entry. *J Am Med Inform Assoc.* 2003;10:188–200.

37. Luff P, Heath C, Greatbatch D. Tasks-in-interaction: paper and screen-based documentation in collaborative activity. In: Turner J, Kraus R, Eds. *Proceedings of the Conference on Computer Supported Cooperative Work.* New York: ACM Press; 1992:163–170.

38. Coiera E, Jayasuriya R, Hardy J, Bannan A, Thorpe MEC. Communication loads on clinical staff in the emergency department. *Med J Australia.* 2002;176: 415–418.

39. Tellioglu H, Wagner I. Work practices surrounding PACS: the politics of space in hospitals. *Computer Supported Cooperative Work.* 2001;10:163–188.

40. Patel VL, Kushniruk AW. Understanding, navigating, and communicating knowledge: issues and challenges. *Meth Inform Med.* 1998;37:460–470.

41. Garfinkel H. *Studies in Ethnomethodology.* Englewood Cliffs, NJ: Prentice-Hall; 1967.

42. Garrod S. How groups coordinate their concepts and terminology: implications for medical informatics. *Meth Inform Med.* 1998;37:471–476.

43. Hutchins E. *Cognition in the Wild.* Cambridge, MA: MIT Press; 1995.

44. Berg M. Practices of reading and writing: the constitutive role of the patient record in medical work. *Sociol Health Illness.* 1996;18:499–524.

45. Faber M. Design and introduction of an electronic patient record: how to involve users? *Meth Inform Med.* 2003;42(4):371–375.

46. Suchman L. Working relations of technology production and use. *Computer Supported Cooperative Work.* 1994;2:21–39.

47. Siddiqi J, Shekaran MC. Requirements engineering: the emerging wisdom. *IEEE Software.* 1996;13:15–19.

48. Reddy M, Pratt W, Dourish P, Shabot MM. Sociotechnical requirements analysis for clinical systems. *Meth Inform Med.* 2003;42:437–444.

49. Star SL (ed.). *The Cultures of Computing.* Oxford, UK: Blackwell; 1995.

50. Zuboff S. *In the Age of the Smart Machine: The Future of Work and Power.* New York: Basic Books; 1988.

51. Panzarasa S, Madde S, Quaglini S, Pistarini C, Stefanelli C. Evidence-based careflow management systems: the case of post-stroke rehabilitation. *J Biomed Inform.* 2002;35:123–139.

52. Collins HM. *Artificial Experts: Social Knowledge and Intelligent Machines.* Cambridge, MA: MIT Press; 1990.

53. Berg M. Search for synergy: interrelating medical work and patient care information systems. *Meth Inform Med.* 2003;42:337–344.

54. Hughes D. When nurse knows best: some aspects of nurse/doctor interaction in a casualty department. *Sociol Health Illness.* 1988;10:1–22.

55. Goorman E, Berg M. Modelling nursing activities: electronic patient records and their discontents. *Nurs Inquiry.* 2000;7:3–9.

56. Gasser L. The integration of computing and routine work. *ACM Trans Office Inform Sys.* 1986;4:205–225.

57. Schmidt K, Bannon L. Taking CSCW seriously: supporting articulation work. *Computer Supported Cooperative Work.* 1992;1:7–40.

58. Gorman PN, Lavelle M, Ash JS. Order creation and communication in health care. *Meth Inform Med.* 2003;42:376–384.

59. Kilpatrick ES, Holding S. Use of computer terminals on wards to access emergency test results: a retrospective audit. *Brit Med J.* 2001;322:1101–1103.

60. Suchman L. *Plans and Situated Actions: The Problem of Human–Machine Communication.* Cambridge, UK: Cambridge University Press; 1987.

61. Kay S, Purves IN. Medical records and other stories: a narratological framework. *Meth Inform Med.* 1996;35:72–87.

62. Bardram J. Temporal coordination: on time and coordination of collaborative activities at a surgical department. *Computer Supported Cooperative Work.* 2000;9: 157–187.

63. Rouncefield MF, Hartswood M, Proctor R, et al. Making a case in medical work: implications for the electronic medical record. *Computer Supported Cooperative Work.* 2003;12:241–266.

64. Svenningsen S. *Electronic Patient Records and Medical Practice: Reorganization of Roles, Responsibilities, and Risks.* Copenhagen: Copenhagen Business School Thesis; 2002.

65. Kling R, Ed. *Computerization and Controversy: Value Conflicts and Social Choices.* San Diego: Academic Press; 1996.

66. Turkle S. *The Second Self: The Human Spirit in a Computer Culture.* New York: Simon & Schuster; 1984.

67. Leveson NG, Turner CS. An investigation of the Therac-25 accidents. *Computer.* July 1993;18–41.

68. Tsai TL, Fridsma DB, Gatti G. Computer decision support as a source of interpretation error: the case of electrocardiograms. *J Am Med Inform Assoc.* 2003;10:478–483.
69. Brown JS, Duguid P. *The Social Life of Information.* Cambridge, MA: Harvard Business School Press; 2000.
70. Lave J. *Cognition in Practice.* Cambridge, UK: Cambridge University Press; 1988.
71. Ash J, Berg M. Report of conference track 4: socio-technical issues of HIS. *Meth Inform Med.* 2003;69:305–306.

7 Clinical Decision-Support Systems

Based on: Ash JS, Sittig DF, Campbell EM, Guappone KP, Dykstra RH. Some unintended consequences of clinical decision support systems. *Amer Med Informatics Assoc Fall Symposium.* 2007;16–30.

7.1 Key Points

- Analysis of examples of unintended consequences of clinical decision-support systems (CDS) uncovered three themes related to CDS content: elimination or shifting of human roles, difficulty in keeping content current, and inappropriate content.
- Three additional themes related to CDS presentation were found: rigidity of the system, alert fatigue, and potential for errors.
- Management of CDS must include careful selection and maintenance of content and prudent decision making about human–computer interaction opportunities.

7.2 Introduction

Clinical decision-support systems (CDS), broadly defined here as computer-based systems offering "passive and active referential information as well as reminders, alerts, and guidelines,"[1] are an important component of computerized physician or provider order entry (CPOE), meaning direct entry of orders via computer by physicians or others with the same privileges. In fact, CPOE alone may offer little benefit without CDS.[2,3] Together, CPOE and CDS can decrease medical errors[4,5] and improve hospital efficiency[6] and practitioner performance.[7] CPOE can also generate unintended adverse consequences (UACs),[8–10] which the authors have studied for the past three years. One clear pattern across all UAC types that arose during this analysis was that many of the unintended consequences were related to CDS. To answer the question "What are the unintended consequences of CDS?" we conducted a detailed analysis of all CDS-related examples.

7.3 Results

We found two major patterns, with unintended consequences being generated either by (1) the actual content of the decision support module or (2) the presentation of the information on the computer screen.

105

7.3.1 CDS Unintended Consequences Related to Content

The examples related to content were grouped around three themes: the elimination or changing roles of clinicians and staff, the currency of the CDS content, and wrong or misleading CDS content.

Elimination or Shifting of Human Roles

Prior to the emergence of CPOE, clerical staff, pharmacists, and nurses often double-checked physicians' orders. CDS content is sometimes designed to eliminate the perceived need for such verification. The ability of the system to provide assistance with scheduling orders is sometimes suboptimal, whereas prior to CPOE, clerks had helped in this regard—by monitoring X-ray orders, for example. One interviewee noted:

> We probably underestimated initially the gatekeeper function that the clerical staff [performed] questioning daily X-ray orders after a certain amount of time . . . once we automated, chest X-ray orders went on ad infinitum.

Medication ordering is also an issue with CDS. When the nurse or the pharmacist is taken out of the loop and the physician is required to manage the dosing with inadequate CDS, problems can arise. As noted in field notes:

> Glitches with IV drips. Instead of ordering the dose per time you have to order it in drops per time, which means you have to know what the quantity of IV solution it is in, and the doctors then have to manually calculate the drip rate.

Similarly, when ordering is done from locations away from the patient, with CDS but without full information about the patient and available assistance from other staff, problems arise. An interviewee stated:

> The doctors have to write the orders into the computer, and the RT isn't always there beside them, so they are either asking the nurses what the settings are, or they have to go find RT or go check the ventilator themselves.

Currency of the CDS Content

Updating CDS content becomes necessary either because outside influences such as the Centers for Medicare and Medicaid Services (CMS) and the Joint Commission (formerly the Joint Commission on Accreditation of Healthcare Organizations [JCAHO]) mandate it, or because new knowledge becomes available. Even the excellent organizations we studied had difficulty keeping up with CDS changes. Coding for billing or compliance and difficulties updating order sets and rules caused problems. One expert described a coding update issue:

> CMS codes change periodically . . . so if you're using a code whose definition changes because of the government, you're in trouble. On day one, it was seven tests; on day two, it was eight tests because the government redefined it.

Updating to comply with JCAHO requirements can also be a formidable task. One clinician noted:

> There are apparently 4000 places where "cc" is used instead of "ml" [the abbreviation now recommended by JCAHO]. I can't imagine how much work it is going to take to review all of the screens to find them, and what the incidence of new error might be during the fix.

Development of order sets and building local knowledge into the system can make CDS more acceptable, but updating of those items remains problematic. An expert stated:

> These proliferating practices of order sets are out of control. I mean it was beneficial at first . . . [but] they don't manage [update] them personally.

Updating the algorithm-based rules behind the CDS is equally difficult:

> He notes that the references for the rules are often out of date and become a bone of contention, particularly when the staff doesn't want the rule.

Wrong or Misleading CDS Content

This broad category includes practical issues such as the introduction of new CDS modules that encourage ordering even when the hospital does not have adequate supplies, alerts that are inappropriate, and information that is not trusted.

If CDS leads clinicians to order something that is not adequately stocked, there can be problems. For example, one expert stated:

> When reminders are introduced to remind people about doing hemoccult tests or pneumococcal vaccine or whatever, . . . you need to make sure that the inventories of those supplies are adequate because you very quickly run out of those things.

Alert content can be problematic. Inconsequential alerts are especially annoying. As one physician said, "Ninety-five percent of the alerts that were generated to physicians turned out to be inconsequential to the patient—and that's a problem."

Sometimes clinicians do not agree with alerts. Field notes taken during the study noted:

> They would not turn on the drug allergy alerts until they were assured that there was a way to filter out specific allergy alerts that they did not agree with.

Inappropriate alerts are also frustrating. As one resident said:

> For example, the alert to not use broad-spectrum antibiotics such as vancomycin is not appropriate in ICU . . . that's why patients came to the ICU, to get the vancomycin.

Contradictory advice offered by alerts can be confusing, as in the case where an alert warned the user against ordering something, but the system (perhaps through an order set) suggested that it be ordered:

> Hey, you're forcing me to place an order—a replacement order—that I may have seen an alert about that made me not want to place it.

And sometimes decision support on the computer is simply inadequate. As one physician said:

> The online formulary kills me. It is much easier to work from a book . . . I may just want to know how fast to push a med, and the online pharmacy only shows me the pharmacology of the med.

There are sometimes data quality problems that cause clinicians to mistrust information. Medication reconciliation is often an issue:

> Our list of medications is actually a list of medications dispensed, which the patient may or may not be taking at the time you see them.

There can also be mistrust about the provenance of the data. Field notes taken during the study said, "If the source for an allergy is [hospital] X, they will not accept the allergy as being valid until they verify it independently." And one resident said about a cost reminder:

> I know this lab costs more [than what the clinical decision support says], so I just ignore the other costs of the labs.

7.3.2 CDS Unintended Consequences Related to Presentation

Many unintended consequences of CDS stem from the way alerts and other modes of decision support are presented to the user. These appear to be caused by the rigidity of systems, sources of alert fatigue, and sources of potential errors.

Rigidity of Systems

The balance between the need for a system to gather and use structured data and the need for clinicians to be able to work easily and quickly is sometimes upset by CDS. Physicians have been known to use workarounds to avoid spending time entering required data. Our field notes provide an example: "An alert . . . required a numerical entry, and the physicians were just putting a 1 in." Workflow can be interrupted, as one interviewee noted, "One of the unintended consequences is that you can make the workflow much harder by inserting the computer." Lists arranged a certain way can also be a problem:

> Folks may think there is no dose matching what they want. The lists cannot be sorted.

Finally, linear order sets may not mirror the complex reality of ordering:

> Current order sets are organized in a linear fashion [even though a physician] thought most of the problems were multidimensional.

Alert Fatigue

We found more examples of alert fatigue, when clinicians feel that there are too many alerts, than any other aspect of CDS. Following are just six of many examples from our study:

- Drug–drug interactions: Most are ignored.
- Alerts about weight-based dosing—this is a real problem. Weights can't be entered into the system, so this is difficult to override. IT can't seem to disable this [feature] either.
- We over-alert. We either need to put teeth into our warnings or don't do them at all.
- If you d/c a lab test that is part of an order set, a pop-up appears letting you know it is part of an order set.
- Because of so many alerts and pop-ups, the doctors may blithely click OK without a reason and not even read [the message].
- It seems like nearly every drug has an alert associated with it.

Sources of Potential Errors

Many of the problems outlined earlier can lead to errors, but we found several additional situations that seem to be particularly prone to producing errors. Although often helpful and time-saving, auto-complete features can sometimes be problematic. An interviewee noted:

We have synonyms for common misspellings. The doctor puts that in, and an order goes into the patient's chart with the kind of accepted name. [The physician will] come back later and look at that and say, "I didn't order that."

Timing is often a problem, with alerts sometimes being seen when it is too late for action, or leading to delayed action. Field notes stated:

X was never notified the lab test never happened, [nor was] the ER, and the patient had to stay in the hospital overnight. This greatly increased the cost of the hospital stay.

Updating can lead to typing errors. According to our field notes:

I cannot imagine what the incidence of new error might be during the fix, such as eliminating an element in a pick list by accident or making a typo in some drug name.

Several subjects pointed out that clinicians might not pay attention to important alerts because of over-alerting. Two different informants called it "crying wolf":

I want systems to "quit crying wolf.

He is concerned that so many [alerts] will create a "cry wolf" phenomenon where important alerts may be ignored. It is especially prevalent with drug–drug interactions. (From field notes)

7.4 Discussion

The unintended consequences that arise because of CDS can, of course, be avoided if little CDS content is used. Without CDS, however, the major benefits of CPOE cannot be realized. The sites we studied that had locally developed CDS had more issues with content and fewer problems with presentation than did the sites with commercial systems. This is most likely either because the hospitals with commercial systems have a minimal amount of content or because their users have very little control over the presentation specifics within these systems.

The hospitals with home-grown systems that we studied tended to have more CDS content because of the way in which their CDS content was developed. Individual clinicians, who were often the developers of the system, recognized needs and built the CDS content little by little over time. Modules were added gradually, usually after considerable testing.

The hospitals we studied that had commercial systems had less CDS content. Sites with such systems tend to lack CDS content for several reasons. First, the vendors may not have the desired mechanisms for evaluating or displaying the clinical content in the way clinicians want it. Second, community hospitals that are not also teaching hospitals may not have time for and interest in creating new CDS content. Third, the ability to customize content is limited in commercial systems, even though hospitals may want extensive customization. Finally, both the hospital and the vendors may believe that legal issues could arise if content is customized.[11]

It appears that presentation of CDS is a problem for all hospitals, regardless of whether they have locally developed or commercial systems. A good part of this difficulty undoubtedly derives from the natural tension between the need to notify the clinician before he or she makes a mistake and the inability of the computer to know the thought process of the clinician. In addition, the requirements of these systems to collect and use structured data and the ability for the system to allow flexible data entry methods are in direct conflict. When the system is designed so that the clinician must enter data into a certain field or must give an explanation, using a menu, for overriding decision support, the system is collecting important data. However, the clinician may resent this intrusion on his or her time and may perceive that the system is questioning the clinician's judgment.

One limitation of our study is that our sites were selected because they are successful: It is likely that the problems we discovered are more severe elsewhere. Another is that we did not set out to study CDS, although its emergence as a major underlying cause of CPOE-related unintended consequences emphasizes its importance.

We have a number of recommendations for managing the CDS issues outlined in this chapter. Those related to content can be addressed by having a knowledge management (KM) structure in place. KM is an organized method for the selection, development or customization, organization, and maintenance of CDS modules. Some excellent suggestions for organizing KM have been published elsewhere.[12-14] If issues arise because human roles are shifted, the KM plan can assure that staff besides physicians are involved in the choice and implementation of CDS so that workflow is considered and needed double-checking continues. The currency of CDS can also be assured when a KM structure outlines methods for acquiring new knowledge and regularly reviewing and updating each module. Constant updating is desired and inevitable, and a thorough review process before a module goes live is necessary.

Content that is thought to be "bad" for any reason can be addressed through a KM structure by having a process in place for clinicians to easily provide feedback. Problems with the quality of the data entered into the system are serious, because CDS cannot accurately respond to patient data in the system if data are erroneous.[15] These problems can also be addressed through the KM plan if continuous training of users includes not only making them aware of new CDS, but also increasing their awareness about why the data are needed and why users need to be careful when entering structured data. A sound organizational structure for KM must include clinicians from different specialties and departments within the organization. In addition, it is imperative that a process be established for periodically measuring and reviewing key CDS performance metrics.

There are some possible remedies for the presentation issues we uncovered. The rigidity of the systems is often caused by the need to capture structured data. However, if this information is not necessary and the data collection process is interfering with clinician workflow, this step can and should be reconsidered. Alert fatigue can be mitigated by reducing the number of alerts. This can be done by carefully selecting a set of alerts that can have the greatest impact[16] and, if it is possible with the system, by filtering alerts based on the severity levels of interactions.[11] Unfortunately, the ability to modify CDS is highly constrained in many commercial systems. Clinical information systems vendors need to work closely with their customers to better understand CDS needs. Finally, there is need for research about CDS use in hospitals using commercial systems so that the lessons being learned now by those on the forefront can be passed to others.

7.5 Conclusion

During our study of the unintended consequences of CPOE, we found 47 examples of consequences related to CDS. After performing a detailed analysis, we found that these issues were related either to the content of the CDS or to the presentation. While these unintended consequences could be avoided completely if no CDS is implemented, without CDS, CPOE cannot offer the benefits that can lead to safety improvements. Incorporation of patient-specific clinical decision support within CPOE systems is critical for successfully addressing the many challenges facing the modern healthcare industry.

Questions to ask regarding CDS issues in relation to CPOE include the following:

1. Is "a reduction in unnecessary variation in clinical practice" a goal of the organization?
2. Are order sets available for the top 100 admitting diagnoses for the organization?
3. Is periodic (yearly) review of all CDS content carried out by the responsible author?

4. Does the CDS oversight committee include clinical representation?

5. Does the CDS provide access to Internet-based information resources (e.g., Up-to-Date, Isabel, Micromedex)?

6. Is there periodic (e.g., quarterly, semi-annual, annual) monitoring and review of the clinical alert (e.g., pop-up alerts) override rate?

7. Is there periodic (e.g., quarterly, semi-annual, annual) monitoring and review of the percentage of all order sets that are actually used?

8. Is there periodic (e.g., quarterly, semi-annual, annual) monitoring and review of the percentage of all clinical alerts and reminders that actually fire?

9. Is there periodic (e.g., quarterly, semi-annual, annual) monitoring and review of the use of all CDS applications that are not already specified (e.g., antibiotic ordering assistant)?

10. Is there a documented process for identifying and prioritizing new clinical alerts?

11. Is there periodic (e.g., quarterly, semi-annual, annual) monitoring and review of the percentage of all medication doses that are above the recommended maximum dosage?

12. Is there periodic (e.g., quarterly, semi-annual, annual) monitoring and review of the standard deviation of all drug doses?

13. Does the system provide an alert or other information, or is the order blocked, when the user orders medications that result in therapeutic overlap with another new or active order (may be same drug, same drug class, or components of combination products)?

14. Does the system provide an alert or other information, or is the order blocked, when the user orders a medication in which the specified dose exceeds recommended dose ranges, the specified dose will result in a cumulative dose that exceeds recommended ranges, or the specified dose violates dose limits for each component of a combination product?

15. Does the system provide an alert or other information, or is the order blocked, when the user orders a medication to which an allergy has been documented or allergy to other drug in same category exists?

16. Does the system provide an alert or other information, or is the order blocked, when the user orders a medication and specifies a route of administration that is not appropriate for the identified medication?

17. Does the system provide an alert or other information, or is the order blocked, when the user orders a medication that results in a known dangerous interaction when administered together with a different medication or results in an interaction in combination with a drug or food group?

18. Does the system provide an alert or other information, or is the order blocked, when the user orders a medication that is contraindicated based on patient diagnosis or a diagnosis that affects the recommended dosing?

19. Does the system provide an alert or other information, or is the order blocked, when the user orders a medication that is contraindicated based on patient age or weight?

20. Does the system provide an alert or other information, or is the order blocked, when the user orders a medication that is contraindicated based on laboratory studies or for which laboratory study results must be considered for dosing?

21. Does the system provide an alert or other information, or is the order blocked, when the user orders a medication that is contraindicated for this patient based on interaction with contrast medium (in ordered radiology study)?

22. Does the system provide an alert or other information, or is the order blocked, when the user orders a medication that requires an associated or secondary order to meet the standard of care (prompt to order drug levels during medication ordering)?

23. Does the system provide an alert or other information, or is the order blocked, when the user orders a test that duplicates a service within a time frame in which there are typically minimal benefits from repeating the test?

24. Does the system *not* provide an alert or other information when the user orders a medication that will produce such a slight or inconsequential interaction that clinicians typically ignore the advice/prompt?

7.6 References

1. Bates DW, Kuperman GJ, Wang S, et al. Ten commandments for effective clinical decision support: making the practice of evidence-based medicine a reality. *J Am Med Inform Assoc.* 2003;10(6):523–530.

2. Nebeker JR, Hoffman JM, Weir CR, Bennett CL, Hurdle J. High rates of adverse drug events in a highly computerized hospital. *Arch Intern Med.* 2005;165(10):1111–1116.

3. Han YY, Carcillo JA, Venkataraman ST, et al. Unexpected increased mortality after implementation of a commercially sold computerized physician order-entry system. *Pediatrics.* 2005;116(6):1506–1512.

4. Kaushal R, Shojania KG, Bates DW. Effects of computerized physician order entry and clinical decision-support systems on medication safety: a systematic review. *Arch Intern Med.* 2003;163(12):1409–1416.

5. Bates DW, Leape LL, Cullen DJ, et al. Effect of computerized physician order entry and a team intervention on prevention of serious medication errors. *JAMA.* 1998;280:1216–1311.

6. Chaudhry B, Wang J, Wu S, et al. Systematic review: impact of health information technology on quality, efficiency, and costs of medical care. *Ann Intern Med.* 2006;144(10):742–752.

7. Garg AX, Adhikari NK, McDonald H, et al. Effects of computerized clinical decision-support systems on practitioner performance and patient outcomes: a systematic review. *JAMA.* 2005;293(10):1223–1238.

8. Ash JS, Berg M, Coiera E. Some unintended consequences of information technology in health care: the nature of patient care information system related errors. *J Am Med Inform Assoc.* 2004;11(2):104–112.

9. Campbell E, Sittig DF, Ash JS, Guappone K, Dykstra R. Types of unintended consequences related to computerized provider order entry. *J Am Med Inform Assoc.* 2006;13(5):547–556.

10. Koppel R, Metlay JP, Cohen A, et al. Role of computerized physician order-entry systems in facilitating medication errors. *JAMA.* 2005;293(10):1197–1203.

11. Kuperman GJ, Reichley RM, Bailey TC. Using commercial knowledge bases for clinical decision support: opportunities, hurdles, and recommendations. *J Am Med Inform Assoc.* 2006;13(4):369–371.

12. Osheroff JA, Pifer EA, Teich JM, Sittig DF, Jenders RA. *Improving Outcomes with Clinical Decision Support: An Implementer's Guide.* Chicago, IL, HIMSS; 2005.

13. Osheroff JA, Teich JM, Middleton BF, Steen EB, Wright A, Detmer DE. *A Roadmap for National Action on Clinical Decision Support.* Bethesda, MD: American Medical Informatics Association; June 13, 2006.

14. Greenes RA, Ed. *Clinical Decision Support: The Road Ahead.* New York: Elsevier; 2007.

15. Berner ES, Kasiraman RK, Yu F, Ray MN, Houston TK. Data quality in the outpatient setting: impact on clinical decision-support systems. Proc *AMIA Symp.* 2006;41–45.

16. Shah NR, Seger AC, Seger DL, et al. Improving acceptance of computerized prescribing alerts in ambulatory care. *J Am Med Inform Assoc.* 2006;13(1):5–11.

8 Clinical Information Systems and Communication: Reciprocal Impacts

Based on: Dykstra R. Computerized physician order entry and communication: reciprocal impacts. *Proc AMIA Symp.* 2002;230–234.

8.1 Key Points

- Entering orders directly into the computer can result in an "illusion of communication," or the assumption that the order will be received and acted upon by the proper people.

8.2 Introduction

There is substantial literature on computerized physician order entry (CPOE) and error reduction.[1–3] Substantially less has been written about CPOE's effects on work processes, however, even though its impact on these processes can be critical.[4,5] There are also few descriptions of CPOE's effects on established clinical communication processes,[6,7] and fewer still of its effects on communication during implementation.[8]

The purposes of the study described in this chapter were twofold: to examine the impact of CPOE on communication in organizations and to consider how those impacts affect the implementation of CPOE. Virtually all dimensions of communication were affected, from interpersonal communication to broad intra-institutional communication.

8.3 Key Findings: Reciprocal Impacts of CPOE on Communication

8.3.1 An Illusion of Communication

For all involved in patient care, the introduction of CPOE presents a fundamental upheaval. There are obvious changes like electronic transfer of orders to the pharmacy. There are also less obvious yet important changes. For example, the process becomes less tangible. As one study participant noted, "[Doctors] picture it going into a black box, that it's a magic process, it's an illusion of communication." There is also a sense of near-infallibility:

> The computer gives a false sense that communication is happening. You enter it and think it went to the right place. With handwriting, you knew it hasn't gone anywhere, so you pick up the phone.

Over-reliance on system dependability leads to potentially dangerous assumptions:

> I'll occasionally call . . . Otherwise, I'll just assume that it's being done.

8.3.2 A Substitute for Interpersonal Communication

Until the widespread adoption of electronic text communication and CPOE, little exchange of plans, ideas, or orders could occur without some form of fairly direct human-to-human contact. Now that face-to-face interaction is no longer the mainstay, the "personal" nature of communication has diminished:

> I think people now . . . substitute interaction with the computer for communication with individuals. The end result is, "Oh, put it in the computer" rather than "Tell me about it."

People need communication and interaction to understand roles and establish self-worth. A team, after all, is more than the sum of individual roles: "[We] began to look at primary care as a team activity several years ago and tried to create an environment where people could work together." For this reason, marginalizing team cohesiveness can have serious consequences:

> The nursing staff . . . it's affected their espirit de corps . . . because you start doing physician order entry and direct entry of notes, and you move that away from the ward into a room, and now you eliminate the sense of team, and the kind of human communication that really was essential . . . You create physical separation.

8.4 Disturbance of Doctor–Nurse Communication: "There's Not That Physical Presence"

Although the modern medical team has multiple members from pharmacy, physical therapy, nutrition, social work, and other areas, the doctor–nurse dyad remains central in the modern team but is affected by CPOE:

> One of the complaints we've heard is that there's not that physical presence. That people aren't around as much to ask questions to and get this interaction with.

> There's a negative feeling about nurse–doctor communication, and nurses feel it's lost now.

Although inadequate communication could lead to error, more often it leads to additional effort to avoid error: "calling people for small things." Provider remoteness has other implications for patient care: "It's more difficult to get residents to see patients who . . . need to be seen, as the residents can put in orders from anywhere."

Physicians feel the separation as well as nurses:

> [W]e're stuck at the computer all day long . . . entering words, communicating through the computer . . . the personal communication is worse—you know, actually speaking with the nurse.

> There really doesn't seem to be any real interaction with the nurses during rounds.

The significance of these channels of communication is clearly recognized by the participants:

> You know, I think that would just be helpful if, when you're talking about the patient, the nurse is present. They can add things, and they can hear what the plan is for the day.

As clinical staff become accustomed to the new situation, communication strategies change and greater efforts are made:

> I think a little bit extra effort on the part of [physician staff] and a little extra effort on the part of the nurse [are needed] to make sure that we're touching base at least once a day as part of our routine . . .

8.4.1 Impact on Medication Orders: A Three-Way Communication

Medications, laboratory, and imaging orders present additional steps in the process with the pharmacy, laboratory, or X-ray department acting as an intermediary. The pharmacist, for example, can be key to success: "They know the meds better. They are the fail-safe mechanism." For the pharmacy, illegible handwriting was the problem in the past with medication orders, but now the key issues are interpretation and rechecking:

> [A]ll of a sudden we have a situation where, potentially, physicians are entering the wrong drug . . . and then it becomes our responsibility to make sure that the patient is getting the right medication.

Communication volume may increase dramatically initially, largely due to the increasing need for rework:

> I was struck by how many conversations [the physician] had by phone about a drug order—[the physician] hadn't had the time or taken the time to speak with the charge nurse or the pharmacist when s/he put the order in.

Verbal orders have long been problematic in many respects, including in relation to the need for accurate recording and co-signatures. CPOE certainly has not solved these problems but may present a new imperative to face them. "Verbals" have never been popular (except perhaps with physicians):

We did it years ago, too—we would discourage verbal orders. It's always better if it's written by the originator, for clarity and to make sure that everything is exactly as it should be.

8.4.2 Medical Care Team–Patient Communication

CPOE has altered information flow from the medical care team to the patient as well. Encounter information is usually available immediately after the visit and remains so indefinitely. Team members can use this information later to remind patients of plans:

> [In] the outpatient area, patients frequently leave their provider visit with questions they haven't asked. They call back later and . . . the pharmacist or the nurse, can . . . say, "Well, you know, [the doctor] wanted you to do this and this."

Patients commonly seek care from several primary care physicians, naturopaths, and specialists/subspecialists, leading to the situation called "co-management." Usually these clinicians have little or no contact with one another and do not coordinate their care:

> [M]any people are co-managed. They don't tell us about the physicians that they're going to, and they don't tell that physician they're coming here, and so you have the chance that people would get in trouble.

Nonverbal communication, such as eye contact with patients during the encounter, is also important to clinicians and presumably to patients:

> [P]atients will come in and they'll sit in the chair that's really behind you. So you can't look at the computer and the patient. And I always make the patient get up and sit so that I can kind of keep my eye on both of them simultaneously.

8.4.3 Some Communication Improves: Medical Teams

CPOE initially affects communication negatively in some areas, even as other areas may benefit from increased efficiency and support. For example, house staff teams in a teaching hospital may find team closeness productive:

> I always come back to the team room. It's easier, quieter, and I can keep up with other work at the same time. The presence of the rest of the team and the availability of [the resident] is also a factor.

8.4.4 Problems Vary by Time and Location

It is not surprising that the varying configurations of the healthcare team throughout the day and throughout the institution influence the effects wrought by CPOE. First, consider evening and night shifts of nurses and physicians:

It's not that it doesn't happen during the day, because it still does, but it's more likely to happen during the off-hours [that] the physician is not in the unit or out admitting another patient and . . . [writes] an order, and 3 hours later the nurse finds out, because . . . she or he just happens to go in there and check orders.

Second, some units, such as a bone marrow transplant unit (BMTU), may feel the need to have more control over orders and medications than CPOE currently allows:

BMTU orders are so unconventional that most "conventional" staff has a hard time filling them. If orders are not entered and complete by 1800 [hours], the 0600 orders probably won't happen. This is a disaster in a protocol that is orchestrated to the hour.

BMTU orders need to be double or triple co-signed [and the CPOE system] does not allow for that.

Ambulatory care units (ACU) were created to allow for certain brief procedures that are performed most efficiently in a hospital setting to be done without actual admission. Workflow in these units differs in that clinicians write the orders at their convenience but the patient comes in for treatment at the patient's convenience, thereby separating the acts of order writing and order completion:

In the [ACU], we hardly ever see an MD.

[E]verything is paper, except lab reporting; [order entry] from surgery is all paper, meds [are] administered from ward stock.

The Unique Problems of "Hectic Environments"

Hectic environments, for the purposes of this chapter, include the emergency room (ER), intensive care units (ICU), and "code" or "advanced life support" (ALS) situations. In contrast to outpatient or traditional inpatient units, which are somewhat time dependent, these environments are maximally time dependent. Often there is barely enough time to make decisions, let alone document and enter that information in a CPOE system: "[For] emergencies, they're yelling out orders and we're following them." After the crisis, participants enter orders and document the care delivered.

Because ICU care is expensive (often two to three times the regular bed charge) and beds are limited in most facilities, there is pressure to stabilize and transfer patients as soon as possible. This means carrying out most orders as soon as possible, which may be more difficult with CPOE systems:

[I]n a critical care unit, the physicians, if they're entering new orders, need to verbally communicate that, because . . . a new order could go in, and then I may not get to giving that stat dose of Lasix for 2 more hours because I won't know it's in there.

In the ER, the interposition of CPOE between the person ordering the medication and the person carrying the order out is a problem:

> Orders are written and handed to the nurse who will administer from ward stock. [The lab procedure] is the same since the nurse is the one who will draw the lab anyway.

This close cooperation is needed for timeliness:

> There's no way that a physician writes the orders in the computer and the nurse is immediately alerted . . . as opposed to a piece of paper they can write on and hand to the nurse.

Working together closely can bring about a familiarity and synergy that are difficult to reproduce electronically:

> [The RN] asked, "Do you want to give solumedrol 125 mg?" "Yes" . . . as he entered the order. This to me was evidence that they had worked together for a long time. (From field notes)

Bidirectional Administration–Staff Communication

Leadership values CPOE for the improved economics and enhanced patient safety that it can bring. Clinical staff members, in contrast, often voice concerns about inefficiencies and loss of autonomy as well as patient safety. Having a common vision for the system is a key to its success: "With an implementation like a physician order-entry system, it helps to have that institutional common vision . . . "

Involvement of top management is one of the key themes identified for CPOE success. Management must both communicate its vision to staff and, in turn, listen to staff members' concerns and suggestions. Even the best of top-level administrators may be somewhat out of touch, however:

> I also got the sense that [the administrator is] not really in touch. I asked about doctor–nurse communication and [the administrator] isn't aware of any problems there. When one is on the unit, you can't help but be made aware of it.

What counts is the CEO's support when it is needed:

> Apparently, our CEO . . . is very committed to an electronic medical record and made sure that we [have] all that's needed.

Such support may even be the sustaining factor:

> [W]hat kept us going here during the tough times is the fact that our administration . . . realized the importance of keeping it going . . . give them the help, the support that they need to keep it going.

Multiple information channels—such as electronic mail, "a message or tip of the day" in information systems, newsletters, publications, and department and general meetings—are used in successful implementations:

[The] clinical lab has a bulletin that comes out every month, so if it has something to do with clinical labs, I can put it in that or I can use any one of numerous publications. But the two most frequently [used communication channels] are the mass email and [information available] on the system itself.

In the absence of correct information, participants may generate their own interpretation: "There was a lot of misinformation and wrong information." The absence of communication breeds distrust and disbelief:

The administration's response to all the stuff preceding [a problematic incident] was sort of small, at least from [the staff's] perspective . . . carrots or bones thrown to [the staff] and never really a true coordinated effort . . .

It is important that the staff be heard and that they perceive that they are being heard: "[A] lot of the anger came from being not heard, from feeling that [staff members] weren't being heard." Opening lines of communication early is extremely important because, once issues are out of control, problems may prove intractable:

Everybody would vent and yell and scream at them, and they would try to explain what they were doing. It was always never enough because the issue had already escalated so far out of control that you could never get ahead of it again.

Disastrous situations are not necessarily due to inattention to detail or callous indifference:

But the problem is that we were naive, I think, and we as an institution . . . the groundwork was not properly laid even though we had a quote, pilot, and that sort of thing.

Special people with personal, communication, and teaching skills are needed to ensure successful implementation of CPOE. Those individuals must have credibility with their audiences and keep the staff informed of system changes and policies:

He's a very patient individual; he's a good communicator; he has clinical relevance to the others because he works beside them.

8.5 Discussion

Communication within the multifunctional medical team seems to have suffered the most, whereas resident–physician teams have seen benefits. Physicians, once they are able to enter orders from anywhere, do so, reducing their interpersonal contact with the floor team. This trend has adverse effects on team relationships. Specifically, team spirit and cohesion are undermined. Reduced face-to-face contact leads to rework when members of the team later need clarification of physician orders. Resident–physician teams may actually benefit from sequestration in team rooms where they are able to interact more closely.

"Top-down" communication is also strained by CPOE implementation. Communication channels that were adequate in times of stability may fail during the upheaval of CPOE implementation. All categories of staff need to be kept abreast of changes related to the system. First, they need information simply so that they can do their jobs. Knowing when and how systems will be activated as well as who to contact when those systems malfunction is essential. Second, a lack of information leads to an atmosphere of mistrust and rumor that erodes morale. Acknowledgment of staff contributions to, and sacrifices for, implementation success is indispensable, as is feedback to staff members indicating that their suggestions, complaints, and feelings are being heard, are valued, and will influence events.

Similarly, "bottom-up" feedback to administration members and the implementation team is essential to keep the endeavor on track and on time. Problems must be addressed quickly. Delays are not readily excused by those affected by the problems. Most institutions have individuals who are familiar with and enthusiastic about information technology in medicine who may act as facilitators in the identification and reporting of problems.

8.6 Conclusion

Plan to be surprised! The institutions studied were all surprised and challenged by unforeseen snags—some worse than others. Form a strong, agile support team, whose members are fully prepared to respond to the unexpected challenges that *will* occur. Even worse than not having a quick answer is having a support system that freezes with no answer at all. Develop a network of superusers, champions, and others who report problems quickly. The earlier problems are identified, addressed, and remediated, the less negative impact they will have. Also, work with this early-warning group to develop solutions to problems as they arise.

In the long term, even a suboptimally done implementation will likely succeed given human flexibility and adaptability. Careful preparation and communication can, however, avoid much user (and implementer!) grief. Once issues balloon to the point they are "out of hand," regaining control will be difficult. "Over-communication" early in the process could represent a solid investment in success.

CPOE perturbs the stability of healthcare organizations by replacing known communication channels with new, often poorly understood channels. Trust in the "black box" leads to an "illusion of communication" that may mask errors. Participants may be unaware of communication problems. If implementers engender a climate of caution and "cross-checking" through person-to-person contact, this climate may not only reduce error but also lead to process improvement.

No matter how much an institution prepares for the upheaval associated with adoption of CPOE, understand that it will probably not be enough. One of the institutions studied prioritized remodeling of processes as a pre-implementation strategy but nonetheless had problems. The unanswerable question is obviously, "How bad would it have been without that effort?"

Questions to ask regarding communication issues in relation to CPOE include the following:

1. Is there evidence of communication of system plans to users?

2. What percentage of regularly scheduled clinical information system oversight meetings actually take place?

3. What is the mean percentage of members of the clinical information system oversight committee who have attended each meeting over the last year?

4. Are there regularly scheduled meetings of the clinical information system safety oversight committee? This committee should be composed of both clinicians who are involved with the clinical information system and representatives from the information technology department, along with specialists in clinical informatics.

5. How would you characterize the evidence surrounding the understanding and communication of the benefits of clinical information system usage to the organization?

8.7 References

1. Teich JM, Merchia PR, Schmiz JL, Kuperman GJ, Spurr CD, Bates DW. Effects of computerized physician order entry on prescribing practices. *Arch Intern Med.* 2000;160(18):2741–2747.
2. Shojania KG, Yokoe D, Platt R, Fiskio J, Ma'luf N, Bates DW. Reducing vancomycin use utilizing a computer guideline: results of a randomized controlled trial. *J Am Med Inform Assoc.* 1998;5(6):554–562.
3. Bates DW, Teich JM, Lee J, et al. The impact of computerized physician order entry on medication error prevention. *J Am Med Inform Assoc.* 1999;6(4):313–321.
4. Massaro TA. Introducing physician order entry at a major academic medical center: I. Impact on organizational culture and behavior. *Acad Med.* 1993;68(1):20–25.
5. Weir C, McCarthy C, Gohlinghorst S, Crockett R. Assessing the implementation process. *Proc AMIA Annu Symp.* 2000:908–912.
6. Zimmerman M. Provider order entry: it can work! *J Emerg Nurs.* 1997;23(5): 463–466.
7. Payne TH. The transition to automated practitioner order entry in a teaching hospital: the VA Puget Sound experience. *Proc AMIA Annu Symp.*1999:589–593.
8. Massaro TA. Introducing physician order entry at a major academic medical center: II. Impact on medical education. *Acad Med.* 1993;68(1):25–30.

9 Overdependence on Technology

Based on: Campbell EM, Sittig DF, Guappone KP, Dykstra RH, Ash JS. Overdependence on technology: an unintended adverse consequence of computerized provider order entry. *Am Med Inform Assoc Fall Symp.* 2007;94–98.

9.1 Key Points

- Overdependence on technology can be an important unintended adverse consequence of the automation of patient care with computer-based provider order-entry (CPOE) systems. Awareness of this issue is vital if organizations are to prepare for and effectively deal with system downtime, assure data accuracy, and help clinicians understand that these tools are designed to support clinical judgment rather than replace it.

- System downtime can create chaos when there are insufficient backup systems in place.

- Users often have unreasonable expectations regarding data accuracy and processing.

- Clinicians cannot work efficiently without computerized systems.

- Organizations should develop methods for measuring the overall efficiency of these systems and developing quantifiable strategies for system improvement.

9.2 Introduction

Healthcare delivery has become increasingly dependent on information technology to computerize almost all aspects of patient care, as evidenced by the proliferation of systems—ranging from billing and accounts management to computerized provider order entry (CPOE) to sophisticated image-guided surgery systems.[1] It is not surprising that introducing these systems into the healthcare environment causes shifts in the way healthcare providers perform their work. As both personnel and organizations adapt to these new technologies, unintended adverse consequences may emerge.

The theory of technological determinism holds that technology is the prime force in initiating social change and that the introduction of new technology fundamentally shifts work activities, resulting in transformations of individuals and their social interactions as well as the organizations in which they work.[2] In contrast to this deterministic approach, the theory of social construction of technology posits that technology does not directly shape society; instead, the social context in which the technology is used determines how it is created, diffuses, and becomes part of the organization.[3] Both theories imply that the introduction of technology is associated with significant change; they differ in whether the change is initiated by the technology or the social context in which it is used.

Regardless of the theoretical basis for understanding the change, it is reasonable to assume that some degree of dependence on any technological innovation will inevitably occur if the technology provides users with some perceived, relative advantage over whatever system it supersedes.[4] This reliance is expected and necessary if the technology is to realize the potential for which it is designed. This fundamentally differs from *overdependence* on technology, in which those using technological innovations no longer treat them as flexible tools to *support* work activities, but instead make incorrect assumptions about how these systems work and begin to rely on them, without question or skepticism, to manage critical work activities.

To answer the question "How does the introduction of CPOE create the potential for overdependence on technology in healthcare organizations?" we conducted a detailed analysis of all references to overdependence on technology in our data. The results are presented in this chapter.

9.3 Results

We identified three themes among the unintended adverse consequences related to overdependence on technology. Because some examples were richly descriptive, a single observation or quote could sometimes be coded in two ways. The first (and largest) theme clustered around the problem of practice disruption and loss of patient safety during system unavailability. The second theme highlighted unreasonable expectations related to data accuracy and processing, spanning a range from strong skepticism to potentially inappropriate complete trust in computerized information. The third theme involved the perception that clinicians cannot work without CPOE technology because they cannot keep current on the wealth of clinical knowledge (e.g., drug–drug interactions, clinical guideline recommendations, hospital formulary contents) required to perform their work effectively. Each of these themes is described in this section.

9.3.1 System Downtime

System unavailability, regardless of its primary cause, can "create chaos" for users and organizations. A system is unavailable if users cannot access it, even though it appears to be operational—for example, when the system interface is working but the back-end database is down, when there is insufficient hardware to support workers' needs, or when the system is so slow that work activities cannot be efficiently completed. Such situations can "create a real fight at times to get work done, because [people] are always in need of a computer," according to one study participant. They frustrate busy end users when they incorrectly assume the system is entirely functional, must find workarounds (e.g., leave the unit to find an available terminal) perform redundant work (e.g., document on paper until a computer becomes available), or, in a worst-case scenario, elect to skip the process of documenting important clinical information.

Because hospital systems are so complex, and because they require the careful integration of disparate, specialized software and hardware systems, single-component downtime can greatly interrupt workflow. For example, if the laboratory result reporting system becomes unavailable, clinicians must rely on phone calls and pagers to get results, may lose the benefits otherwise obtained from the display of historical data for trending, and, therefore, may miss important results necessary for optimal clinical care. In addition, failure of a single system component in one area can dramatically affect other areas, leading to cascading effects that may not be anticipated prior to their occurrence:

> They use a whiteboard screen saver in the ER that keeps track of people in the ER. When the hospital registry system goes down, the registry can't provide the [patient's] ID number; it wreaks havoc in the ER.

In this case, failure of the hospital registry system disrupts the functioning of the entire emergency department because no backup system exists to provide temporary registration numbers until the main system can be brought back online.

Complete system downtime, though generally rare, can have disastrous repercussions for clinicians and institutions. One faculty physician summed it up this way: "It's funny now. When the computer goes down, we don't remember how to document on paper." Even if supplemental paper forms for clinical documentation are available during a downtime event, clinicians can be severely hampered by their inability to access medical data that exist only in electronic from. This loss of important historical data can result in potential for medical error because the clinician must work with incomplete information. In addition, poor preparation for downtime events (e.g., no preparatory training exercises, poorly defined downtime procedures, lack of sufficient paper forms to support documentation until the systems begin functioning again) can magnify the negative repercussions of downtime, as clinical staff are left without necessary electronic resources and have few practical alternatives for efficiently completing their work.

9.3.2 Data Accuracy

Clinicians report a variety of attitudes regarding data accuracy with the computer. When assessing the accuracy of "read" data (e.g., data stored in and provided by the computer for viewing), opinions range from strong skepticism to complete trust of what is presented. In one system we observed, the original source of data was listed beside any displayed information imported from an outside hospital. One clinician reported that she did not trust any of the data from the outside source, and would not use the information until she could verify it. In a life-threatening situation, especially if the patient, his or her family, or friends were not available to provide medical history, such lack of trust in the data could be a concern. In marked contrast, others reported, "If it's in a computer, it must be accurate and complete." Such an assumption is also concerning: If clinicians do not pause to question electronic data, they may fail to recognize errors in the record.

When discussing "write" data (e.g., information input into the system via keyboard, mouse, or other means), clinicians appear to make assumptions about what the system can and cannot do with data as it is entered, or how it is ultimately processed. For example:

> [Some doctors] don't understand that the free-text allergy information cannot be used by the decision support system . . . [they] have a false sense of security as a result.

This misunderstanding is common; there is a perception that data, once entered into the computer, are fully accessible and useful, regardless of how or where they were entered. For this reason, clinicians often enter data into miscellaneous fields when the correct location is not readily found. Other providers accessing the record at a later date may not think to look for information in these nonstandard fields, however, so the data may be inadvertently lost even though they have been entered into the system.

Finally, there is a strong tendency to assume that processes will be carried out to completion once they have been electronically initiated. For example, it is common to assume that because a medical order has been entered into and processed by the system, the requested action has actually occurred. This is especially true in systems where the medication administration record is not available online, so clinicians do not have a single electronic resource to review both the medications that have been ordered and the medications that have been administered:

> The difference between the medication order list and the medication administration list causes the physicians to have a false sense of security . . . many physicians assume that all of the medications that have been ordered have been administered.

9.3.3 Inability of Clinicians to Work Without Automated Systems

CPOE systems with embedded clinical decision support (CDS) provide clinicians with a variety of knowledge support tools, such as notification of drug–drug interactions, warnings about allergies, recommendations for clinical guidelines compliance, and more. For CDS to be effective, it must be current, context sensitive, and well-integrated into the CPOE system, so that clinicians can rely on its suggested clinical guidance to supplement knowledge gaps. When these criteria are met, CDS can be enormously effective for supporting clinical practice. Conversely, when clinicians rely on CDS to the exclusion of sound clinical judgment, the potential for errors can increase.

CDS alerts often "fire" when they have not been properly programmed to leverage important and available information. For example, anticoagulants such as heparin are not commonly administered with aspirin. Nevertheless, this drug combination is often ordered intentionally in the coronary care unit for its heart-protective benefits. In this setting, firing the clinical alert to warn the prescribing clinician is likely unnecessary.

But if the alert does fire, and if the clinician relies on the alerting information exclusively, the possibility exists that the clinician might delete one of the two necessary medications for therapy, thereby increasing the potential risk to the patient.

Because many CPOE systems have been in place for decades, some medical students and residents trained at the institutions possessing these systems have never had to practice medicine without the support of computerized systems. These clinicians may be overly dependent on technology—and they may not be able to efficiently work without it. Studies have shown that the presence of CDS does not appear to negatively affect learning, and may actually improve it;[5] however, moving from the highly integrated electronic medical record or CPOE system back to a non-automated system can prove very difficult for the clinician not familiar with a paper-based clinical record:

> We had a resident who was voted the best resident . . . 2 years in a row . . . a wonderful guy. He took a new position at a new hospital . . . and the head of the medical staff called the residency program director about a month after he got there and said, "I just don't understand; this guy is nonfunctional." He didn't know how to work in a place that didn't have order entry or results retrieval. He took almost 6 months to re-acclimate, [and to] figure out how to order in a different environment.

9.4 Discussion and Suggestions for the Future

Reasonable dependence on technology is a desirable outcome of the automation of patient care systems. Clearly, CPOE systems with integrated CDS provide many distinct advantages to the busy clinician who must synthesize and remember an ever-increasing body of clinical knowledge. However, overdependence on technology can arise when computerized clinical systems are not robust (e.g., are slow or are partially or completely unavailable), when clinicians begin to trust these systems without question, and when healthcare workers have no exposure or training in non-automated clinical environments. For these reasons, it is imperative that healthcare organizations consider and prepare for potential problems related to technology adoption.

9.4.1 System Downtime

One hundred percent reliable systems do not exist; downtimes, whether planned or unanticipated, are inevitable and costly. Even a system with 99% uptime potentially might be unavailable for 14.6 hours per month, with the actual number of person-hours of work lost depending on the number of staff affected by the downtime. One study suggests that for each minute of unavailable system time, staff must spend 4.5 minutes to complete the work they would have done had the system been available, and to reenter the data once the system comes back online.[6] The actual costs of this downtime range from approximately $264 per minute of downtime for a 500-bed hospital to as much as $1000 per minute for a three-hospital integrated delivery network (IDN) with 1400 beds. Over a single year, each 1% of downtime could result in an additional $1.4 million in operating costs for the 500-bed hospital and as much as $10 million for the IDN.[6]

Given these high costs, organizations should formally measure the performance of their systems and develop metrics to assess the financial and workflow impacts of system unavailability. At a minimum, organizations should be able to report their uptime percentages, time to recover for different types of system failures, and overall system usage statistics (e.g., percentage of providers entering orders online, average user load, average network speed). In addition to developing and monitoring these system performance measures, healthcare organizations must develop and test contingency plans for continued operations during system downtimes, so that system unavailability is minimally disruptive to healthcare work. Such plans should include scheduled downtimes for system maintenance, preparations for short-term and long-term system outages, and rigorous protection of data against loss. In addition, downtime preparations should detail provisions for workable paper backup systems, procedures for operating in the absence of electronic resources, and training for employees. To protect against data loss, robust and reliable backup systems must be in place, and they should be rigorously tested with periodic "downtime drills" to assure that they function as expected.

9.4.2 Data Accuracy

Clinicians cannot reliably trust information accuracy if the organization does not take steps to assure it. In the optimal practice of clinical care, information accuracy should be as ubiquitous and necessary as handwashing. Data imported into systems from any source should be rigorously validated for accuracy—a step that requires periodic manual checking of results, development of quantifiable benchmarks for data quality, careful attention to variation from established benchmarks, development of procedures to identify and report data inconsistencies, and dissemination of this information to system users and administrators. Data integrity and reliability should be a key organizational goal.

End users of these information systems should be trained in the proper use of clinical applications, including correct data placement on electronic forms, the reason why standardized data entry can improve data reliability and interoperability (and how to use system tools provided for entering these data), and methods for assuring clinical orders are actually completed in a timely fashion. Many of these training areas imply the need for integrated informatics education in medical curricula. In addition, clinicians should be reminded to carefully evaluate and leverage the information provided to them by CDS so that they make informed and clinically relevant decisions regarding patient care, while using this tool as one of many potential sources of information. Finally, clinicians should be given periodic feedback regarding the data they enter in such areas as use of CDS (e.g., how often alerts are ignored), comprehensiveness (e.g., the presence and/or absence of clinically vital data), and quality (e.g., use of nonstandard abbreviations or over-reliance on free-text entry when standardized entry is available), so that they may improve their documentation.

Inability of Clinicians to Work Without Automated Systems

As clinical care becomes increasingly more dependent on automation, we expect clinicians to rely even more heavily on technology. However, great care should be taken to educate clinicians that over-reliance on technology can be dangerous when it is used to the exclusion of sound clinical judgment. No automated system available today can discern and evaluate all of the subtle physical cues displayed by a patient in a clinical encounter. Despite continued improvements in functionality, proven gains in practice efficiency, and improved access to knowledge sources, these systems are not foolproof. The clinician should, therefore, utilize his or her education and experience in combination with these tools to provide optimal care.

Regardless of the system in use, healthcare organizations should find ways to measure how well these systems actually support or enhance clinical work and quality of care. Organizations should create specific, robust, repeatable, and scalable measures of system performance above and beyond the basic "return on investment" calculations. It is vital to know how often systems are unavailable and to understand and prepare for the impact of downtime on the staff who must rely on these systems to do their work. Measures must be developed to evaluate data quality, accuracy, and comprehensiveness, particularly with regard to CDS effectiveness and the potential for overdependence on these tools. Finally, organizations should share benchmarking strategies with other institutions so that organizations can develop common, useful, and repeatable methods for assessing system performance across the entire healthcare spectrum.

9.5 Conclusion

Overdependence on technology can be an important unintended adverse consequence of the automation of patient care with CPOE systems. Awareness of this issue is vital if organizations are to prepare for and effectively deal with system downtime, assure data accuracy, and help clinicians understand that these tools are designed to support clinical judgment rather than replace it. Finally, organizations should develop methods for measuring the overall efficiency of these systems and quantifiable strategies for system improvement.

Questions to ask regarding technology overdependence issues in relation to CPOE include the following:

1. What percentages of workstations on the organization's network have installed and are running the most recent version of the desktop operating system (e.g., Microsoft, Macintosh, Linux)?

2. How many versions behind the manufacturer's latest desktop operating system is the average workstation in the organization (e.g., as of July 2009 Microsoft's Windows 7 was current, and previous versions in order were Windows Vista, Windows XP Professional, Windows 2000, Windows ME, Windows 98, Windows NT, Windows 95, and Windows 3.1)?

3. How many days behind is the organization in upgrading to the CIS vendor's most recent product release?

4. What is the total number of database transaction errors that are occurring on a daily basis?

5. Is there a secure (encrypted data and secure facility), daily, weekly, and monthly, off-site backup of the main clinical database?

6. Is there a record of testing (not less than once per year) the viability (can the backup copy be restored) of the recent (last 12 months) database backups?

7. Does the organization have available emergency power backup systems, fuel for those systems, and evidence of periodic testing?

8. Is there an external "hot site" for clinical applications? The time (hours:minutes) to come online should be recorded during each test. The currency of the clinical and demographic information available at the hot site should be measured and recorded.

9. Is a downtime emergency kit available on every clinical unit? This kit should contain enough copies of all the organization's approved paper order entry and clinical documentation forms and labels to send samples to the laboratory and care for patients for as long as 2 days. It should also contain a copy of the CIS downtime and reactivation procedures. The reactivation procedures should specify who is responsible for back-loading all of the data that piled up during the downtime. In addition, these procedures should include the order (e.g., admitting, then pharmacy, then laboratory results) in which the old data should be entered. Finally, there must be a method of notifying all the appropriate people that the system is ready for routine use.

10. Does the healthcare organization conduct periodic (at least two per year), unannounced, downtime drills that require the operational leaders of every clinical unit to activate both the downtime and reactivation procedures? The amount of time between the beginning of the drill until each unit calls in to verify that it has all the required forms and data available to continue caring for patients should be recorded. Following these drills, a representative from the clinical information system downtime or CIS safety oversight committee should visit each clinical unit and debrief the leaders of the unit to identify potential problems that occurred during the drill.

11. Does each unit have the ability to print downtime patient data and new data entry forms (assuming that the hospital has power)? These downtime reports should be carefully designed with input from experienced clinicians; they should not simply be a "dump" of all patient data. Each of these reports should be clearly labeled with a message such as, "Please discard following the downtime." Particular care should be given to ICU patient reports. Implementation of such a plan means that there needs to be a PC on each unit that is keeping a copy of all patients' data in a relatively up-to-date manner. Each of these downtime PCs should be backed up by a similar PC on another unit. All downtime PCs should be tested on a daily basis by the charge nurse (e.g., print a single patient's downtime report).

12. Has every clinician attended a short CIS-sponsored downtime preparation class? This class should cover the following issues: where the downtime kits are located, how to use the downtime printers, how the system will be reactivated, and which data need to be back-loaded before the system is reactivated, among other things.

13. Is there a "read-only" version of the CIS (that is clearly identified) that can be activated in the event that the database server of the CIS becomes corrupted?

14. Has an organization-wide downtime committee been established that is responsible for creating and maintaining the downtime and reactivation procedures, designing and distributing the downtime kits, and reviewing the organization's performance following all significant planned and unplanned downtimes? This committee should include representatives from the medical staff, nursing, ancillary departments, administration, and information technology.

15. Is there a planned downtime communication procedure? This procedure should notify all users periodically that the system will be going down starting 30 minutes before the actual downtime.

16. Have downtime notification procedures been established stating who needs to be called, beeped, emailed, and so forth, as well as what to do if the phone system is also down?

9.6 References

1. Shortliffe, EH, Perreault LE, Eds. *Medical Informatics: Computer Applications in Health Care and Biomedicine*. 2nd ed. New York: Springer-Verlag; 2001.
2. Heap N, Thomas R, Einon G, Mason R, Mackay H, Eds. *Information Technology and Society*. Thousand Oaks: Sage; 1995.
3. Pinch TJ, Bijker WE. The social construction of facts and artifacts: or how the sociology of science and the sociology of technology might benefit each other. *Social Stud Sci*. 1984;14(3):399–441.
4. Rogers EM. *Diffusion of Innovations*. 5th ed. New York: Free Press; 1998.
5. Kho A, Henderson L, Dressler D, Kripalani S. Use of handheld computers in medical education: a systematic review. *J Gen Intern Med*. 2006;21(5):531–537.
6. Anderson M. The toll of downtime. *Healthcare Informatics Online*. Available at: http://www.healthcare-informatics.com/issues/2002/04/leading.htm. Accessed February 13, 2007.

10 Persistent Paper: The Myth of "Going Paperless"

Based on: Dykstra RH, Ash JS, Campbell E, Sittig DF, Guappone K, Carpenter J, Richardson J. Persistent paper: the myth of "going paperless." *AMIA Annu Symp Proc 2009. (In press).*

10.1 Key Points

- Analysis of the reasons for the widespread use of paper within all institutions we studied revealed psychological, ergonomic, technological, and regulatory reasons for ongoing use of this documentation medium.
- Paper has many unique attributes, allowing it to fill gaps in information use regarding timeliness, availability, and reliability in pursuit of improved patient care.
- Paper use generally has a positive impact on patient care by serving as a portable, disposable computer interface, supporting information transfer between clinicians and patients, allowing care to go on during outages, and filling gaps in the system to support clinicians.
- As electronic systems assume an archival role, the role of paper is changing from an archival medium to an active and vital communication medium, filling the gaps in current technology.

10.2 Introduction

More than 30 years ago, prognosticators hailed the advent of the "paperless office," suggesting that computer records would completely replace paper in modern businesses. Like the paperless office, however, the paperless hospital has proved to be a myth. The reality is that the number of pages consumed in U.S. offices has been growing by approximately 20% each year.[1] "Going paperless" is often a highly visible goal of implementing computerized provider order entry (CPOE) or an electronic medical record (EMR) system. Shifting to an electronic system rather than a paper-based system is widely expected to improve the efficiency, quality, and safety of medical care.[2-5] Our research team, however, identified "persistent paper"—that is, continued paper use in an electronic environment—as one type of unintended consequence of CPOE implementation.[6]

10.3 Rationale for Persistent Paper

Eliminating paper, thereby creating a paperless office or hospital, was a highly visible organizational objective in all institutions we studied. Despite these noble intentions, however, we observed a great deal of paper still in use in our study. One medium-sized institution (a 300-bed hospital) uses "1.6 million pieces of paper per month—printed or copied . . . [they said] we print and destroy 40% of that paper."

Why the continued use of so much paper in a potentially "paperless" environment? Our analysis revealed four high-level reasons for continued paper use: psychological factors, human and ergonomic factors, technology gaps, and regulatory factors.

10.3.1 Psychological Factors: Paper Is Familiar and Comforting

People and paper have had a long and close relationship. Paper interweaves medical practice and the social fabric of teams. Some people value the palpable presence of paper, needing or wanting a paper document they can see and hold:

> "I like to have a handle on it. I like to have the information on paper where I can hold on to it" (as he clutches his sheaf of papers to his chest).

In our study, participants also stated they found the electronic replacements for comfortable established paper-based processes upsetting:

> I feel at ease if I write [orders] down; I feel like I really know them. In eMAR, you just click on them; it's weird.

Users frequently mentioned difficulties with understanding or processing information in electronic form compared to paper:

> I can see the screen and write. It's not as good but, if I have to go back and forth between screens it takes longer and I make mistakes.

Users did not suggest that information in the paper chart was easier to find or more complete, however. Rather, respondents stated it was easier to understand when information was viewed on paper, so they might print paper copies: "I have to put it down, look at it, and then think about it."

Paper use can support the social systems of care delivery. In the intensive care unit, for example, activity often centers on the patient care flowsheet. As a nurse told us, "It's just the nature of a critical care unit to use a flowsheet." We observed other medical teams organizing work as a group activity. The team divided the work and assigned tasks to individuals who then developed individual to-do lists. Developing the checklist was part of the medical team's social structure and functional hierarchy.

Some continued paper use is due to clinician or institutional "inertia." At several sites in our study, clinicians were provided with a bundle of papers for a clinic visit or an admission but discarded them almost immediately, stating that they really didn't know why the pages were printed since they were no longer used or useful:

> They told us we'd be paperless. Each night all this stuff is printed, [but] we never look at these papers. I don't know why we have them.

10.3.2 Human and Ergonomic Factors: Paper Is Versatile

Ergonomics is the science of refining the design of products to optimize them for human use.[7] Ergonomics also includes a product's existing characteristics such as accessibility and ease of use.[8] "Sticky notes" are a prime example of the ergonomic flexibility observed in all institutions in our study, where they festooned charts, door posts, and computer terminals. Sticky notes held simple requests ("Pt would like order for throat lozenges—Kathy") as well as more complex requests or comments and, of course, computer passwords.

As we noted earlier, everyone from clerks to clinicians used paper reminder lists and check box lists: "On her census pages . . . her patients have check boxes. She crosses them off after she's seen them." These notes and lists often employed color-coding to denote new orders, changes, plans, and other aspects of care.

In the institutions included in our study, we often saw heterogeneous "bundles" of different document types from different sources. For doctors and nurses, as well as for support services, the bundle often consisted of a "census" document with multiple entries for a team or floor, pages for individual patients, and other pages for beepers and contact information:

> They're wonderful to have. If I lose them in the day, I'm totally lost through the rest of the day . . . I have to go all over to get information.

10.3.3 Technology Factors: Paper Fills Gaps

Electronic systems, particularly billing and admission–discharge–transfer (ADT) systems, are widespread, with more than 95% of even critical access hospitals employing them.[9] Utilization of the EMR and CPOE, representing computerization of the care delivery side, is slowly increasing.[10] Because many hospitals and clinics take an incremental approach to implementation, moving bit by bit, unit by unit, to electronic systems, they may bypass some units or functions, resulting in systems that are part paper, part electronic—with paper filling the gaps.

Of course, no computer system can be available 100% of the time:

> [Staff] remembered a time 3 years ago when a server went down and "we had nothing—couldn't get labs or anything . . . no effective paper backup. It was terrible."

Paper backup processes can help an institution avoid grinding to a halt during downtime attributable to either system failure or maintenance. The institutions we observed prepared for such events by maintaining paper copies of recent test results as well as data entry forms including order sets, flowsheets, and lab and X-ray order forms in patient care areas.

As electronic processes replace paper, confusion and duplication often arise:

> We continue to maintain a hybrid documentation environment . . . the following continue to exist wholly or in part as paper: consents, emergency trip records, progress notes . . . nursing assessment and, Code Blue records.

During Code Blue (acute cardiovascular collapse) and other urgent situations, CPOE systems did not replace paper in our study institutions: "[A]ll of that documentation still goes on paper. The paper is then put in the [paper] chart."

Pharmacy routing in some institutions is another place where paper persists:

> A printout is also produced in the inpatient pharmacy with every order entered . . . This process necessitates the use of an unbelievable amount of paper.

Dual documentation can be very time-consuming:

> The transfer note is typed in free text . . . This process is very slow. She prints this out to go with the patient and she calls ahead and gives this report verbally.

Systems may not allow CPOE to be activated for a patient until that individual has inpatient status. Rather than delay delivery of patient care in this circumstance, some institutions support a paper-based admission process that allows paper-based orders to be written, which are later entered into the CPOE system.

We observed hybrid computer–paper instruments in use in several locations in our study. The EMR/CPOE system generated patient lists as well as succinct thumbnail descriptions of a patient's medical condition, pending orders, laboratory results, and other data. Clinicians then wrote more information on them, producing hybrid documents. Where these documents existed, they seemed to be in the pockets of nearly all clinician and nursing staff: "Each of the participants has a bundle of paper. Most have the long rounds, [doctor–patient] lists and chart notes." Participants expressed confidence in these paper reports: "We use these reports to write progress notes. The synopses are really accurate."

Hospital discharges can often be time-consuming. Here, the computer–paper connection can lead to time savings by reducing the time needed for a discharge summary: "If the patient is in for less than a week, it [the EMR] will auto-populate a discharge summary." A timely discharge summary saves time for the discharging doctor and can be a boon for outpatient follow-up care.

10.3.4 Regulatory Factors: Paper Use May Be Required

The regulatory environment can have significant effects on computer and paper use. At one institution, according to a physician informatician, "the official medical record is half computerized, half paper." States may require paper copies of documents such as informed consent, resuscitation status, or advanced health directives to physicians. Narcotic prescriptions often require a "wet signature." The electronic solution in such cases may entail scanning a paper document into the chart as an image.

Institutional policies often require verbal orders be written and placed in the chart if the transcriber cannot enter the order directly into the computer. Sometimes the official order sheets are not available. One participant described the unauthorized use of a hastily grabbed napkin or paper towel as the "paper towel interface," noting, "[Y]ou just can't file that in the chart."

10.4 Discussion

Paper has physical attributes that make it particularly effective in collaborative work. Luff described paper's "tailorability," or rapid customizability that allows, for example, convenient use of various colored markers and freehand annotations.[8] Ecological "flexibility" describes paper's attributes: small, foldable, and easily moved about.[8] In an emergency, paper works without power or access to an information infrastructure. The observed flexibility of sticky notes may, in fact, delay their replacement with electronic equivalents. Just as important, the practice of medicine has developed using paper to communicate. As Harper and Sellen observed, "Paper has helped to shape work practices, and work practices have been designed around the use of paper."[9] While paper's persistence may be inevitable, however, there are good reasons to be cautious about its continued use.

10.4.1 Paper's Positives

Paper's tailorability and flexibility as described by Luff[8] combine to make paper an effective and efficient means of communication in health care. In the sites covered in our study, paper filled procedural gaps, sometimes serving as first choice where speed or improvisation was essential. Where systems do not allow CPOE before the patient arrives, paper could bridge the gap and allow care to begin without delay. There have been instances in which the lack of this intermediary function has caused significant problems.[10] Paper is also needed to communicate health information between electronic and paper-based institutions.

Paper's role during system downtime is similar: When CPOE is unavailable, paper becomes an important backup system to support patient care work. Many institutions have implemented carefully designed procedures detailing how temporary paper records would merge with computer records. Rather than limit the availability of paper documents or the ability to print them, these institutions have worked to preserve paper documents and redefine processes to support continuity of care.

Computers should make graphic presentations easy, yet we saw gaps and lags in developing technology to replace paper-based displays. Flowsheets were the most common symptom: Paper flowsheets the size of newspaper pages displayed monitoring, medication, and intervention information. We also observed that institutions were aware of, and were actively addressing, the graphics problems.

10.4.2 Paper Plus CPOE: Happy or Troubled Marriage?

We saw paper-based memory devices or checklists in all institutions in our study, continuing a longstanding clinician practice and taking advantage of paper's flexibility. As previously described, paper–computer hybrid documents and checklists are highly valued by caregivers in multiple roles.

All is not well in the paper–computer hybrid world, however. Where paper fills a gap or lag, as we saw in our examples from pharmacy, the cross-checking process can increase work for healthcare personnel. Patient transfers between electronic- and paper-based units can require a remarkable amount of paper/computer work to ensure that care continues across the transition. Mixing paper and computing may also pose a threat to patient safety. Participants in our study reported cases in which one clinician entered urgent orders into the CPOE system, while another clinician wrote different orders on paper.

10.4.3 Paper's Negatives: A Reason for Moving to CPOE

Documents with static information lose currency as they age. Locations we visited were concerned about the timeliness of the documents on which they depend: "The nurse states with some it is current, then it ages during the day." Similarly, "versioning" occurs when multiple versions of a paper document exist, leading clinicians to wonder which document has the most up-to-date information: "We feel it's safer to go to the original document in the computer."

Keeping paper-based reference information current is a constant challenge:

"Compatibility Reference Guide" is on the wall—a drug interaction chart. It is a four-year-old version.

Documents are only as good as the persons maintaining them:

It can be a blessing or a curse. If not updated, it looks like the admit note and is of little use; but if it is kept up-to-date, it's great.

Paper documents containing patient information constitute an information security hazard. HIPAA addressed this problem, defining protected health information and mandating how to handle it in exquisite detail. Despite HIPAA, a misplaced patient information sheet or chart may become an untraceable breach of confidentiality—untraceable because paper charts, forms, and notes do not have a secure audit trail of those who have seen the document.

10.4.4 Paper's Replacement

We use the term "inertia" to describe a resistance to change or desire to continue the current process. This kind of paper persistence is the most human of factors: We simply do things as we always have, even though those actions are a vestige of times past.

As digital systems evolve, new capabilities should allow replacement of current paper uses. For example, digital technologies are steadily becoming more paper-like by evolving into smaller, lighter, less expensive, and more portable options. Paper documentation is still required in many locations for advanced directives and narcotic prescriptions. Consent forms for procedures or surgeries often persist in paper form, although electronic replacements are being developed. In addition, federal legislation has encouraged replacement of some "wet signature" requirements.[11] What remains to be done is to operationalize electronic signatures on a large scale in systems, policies, procedures, laws, and bylaws.

Berg points out that work tasks are completed by an interrelation of workers and artifacts together.[12] The artifact (a paper order sheet or CPOE) does not just support the work, but rather, by actually taking part in the work, brings about task completion. Replacement of an element (paper) of this co-developed process is not trivial and will take time and patience. Given enough time, the screen may become as useful as a piece of paper and an electronic signature as forceful as a handwritten signature.

10.4.5 Paper's Future: What Next?

Paper is becoming a transitory medium in healthcare settings: Rather than being archived at the end of the session, it is often simply shredded. This disposal is in line with current industry practice.[13] Some institutions retain little paper; others have no official paper record at all. Where paper continues to be generated for office visits or hospital admissions, it is often shredded at the end of the encounter. In a move that may possibly circumvent the shred–recycle process, Xerox is developing a paper that will erase itself in approximately 16 hours.[13]

Finding ways to accomplish tasks in the new electronic milieu may require adapting old methods to fit the emerging environment. As improved digital devices take on more of the ecological flexibility of paper, we may also expect them to take on more of paper's current roles. Meanwhile, paper will continue to fill many gaps, from psychological to technological to legal.

10.5 Conclusion

We observed paper to have a prominent place, filling vital roles in all institutions we visited as part of our study, with heavy use even at the most computerized of locations. Our observations indicate that "going paperless" is a journey, not an event. As institutions introduce electronic systems, they should not expect the total elimination of paper, nor should they restrict access to paper, the use of paper forms, or the ability to print documents. We found paper use generally had a positive impact on patient care by serving as a portable, disposable computer interface, supporting information transfer between clinicians and patients, allowing care to go on during outages, and filling gaps in the system to support clinicians. Our observations lead us to believe that as electronic systems assume the archival role, the role of paper will continue changing from an archival medium to an active and vital communication medium, filling the gaps in current technology. As those gaps close, paper may go away—but not anytime soon.

Questions to ask regarding "persistent paper" issues in relation to CPOE include the following:

1. What is the ratio of printers to staffed beds for the inpatient setting (sorted by ICU and acute care)?

2. What is the ratio of printers to clinical exam rooms in-use for the out-patient setting?

3. Does the system have the capability to support mobile clinical users? For example, are wireless networks in place and have computers on wheels (COWs) or hand-held, notebook-size computers been deployed?

4. Are recycling bins for paper/printouts containing patient-identifiable data made of metal and locked (e.g., a slotted lid with a locking mechanism)?

10.6 References

1. Abramovitz J, Mattoon A. *Paper Cuts: Recovering the Paper Landscape.* WorldWatch Paper 149. December 1999.
2. Mekhjian HS, Kumar RR, Kuehn L, et al. Immediate benefits realized following implementation of physician order entry at an academic medical center. *J Am Med Inform Assoc.* 2002;9(5):529–539.
3. Kuperman GJ, Gibson RF. Computer physician order entry: benefits, costs, and issues. *Ann Intern Med.* 2003;139(1):31–39.
4. Bates DW, Teich JM, Lee J, et al. The impact of computerized physician order entry on medication error prevention. *J Am Med Inform Assoc.* 1999;6(4): 313–321.
5. Leapfrog Group. Factsheet: Computer physician order entry. 2006. Available at: www.leapfroggroup.org/for_hospitals/leapfrog_safety_practices/cpoe. Accessed September 17, 2006.
6. Campbell EM, Sittig DF, Ash JS, et al. Types of unintended consequences related to computerized provider order entry. *J Am Med Inform Assoc.* 2006;13(5): 547–556.
7. SearchWebServices Definitions. Human factors. 2007. Available at: searchwebservices .techtarget.com/sDefinition/0,290660,sid26_gci214386,00.html. Accessed June 2, 2007.
8. Luff P, Heath C, Greatbatch D. Tasks-in-interaction: paper and screen-based documentation in collaborative activity. In: Turner J, Baecker R, Eds. *Proceedings of the 1992 ACM Conference on Computer Supported Cooperative Work.* New York: ACM Press; 1992:163–170.
9. Harper R, Sellen A. *Paper-Supported Collaborative Work.* Xerox Technical Report EPC-1995-109. 1996.

10. Han YY, Carcillo JA, Venkataraman ST, et al. Unexpected increased mortality after implementation of a commercially sold computerized physician order-entry system. *Pediatrics*. 2005;116(6):1506–1512.

11. Electronic Signatures in Global and National Commerce Act of 2000; ESIGN, Pub. L. No. 106–229, June 30, 2000.

12. Berg M. Accumulating and coordinating: occasions for information technologies in medical work. *Computer Support Coop Work*. 1999;8:373–401.

13. Markoff J. Xerox seeks erasable form of paper for copiers. *New York Times*. November 27, 2006. Available at: www.nytimes.com/2006/11/27/technology/27xerox.html?_r=1&scp=1&sq=&st=nyt. Accessed July 16, 2009.

Overcoming the Unintended Adverse Consequences Associated with Clinical Information Systems

11 Considerations for Successful CPOE Implementations

Based on: Ash JS, Stavri PZ, Kuperman GJ. A consensus statement on considerations for a successful CPOE implementation. *J Am Med Inform Assoc.* 2003;10(3):229–234.

11.1 Introduction

This chapter presents a set of considerations that every organization planning on implementing a clinical information system (CIS) should review prior to and then periodically during its implementation process. During our study, our group realized that guidelines or recipes for success would be difficult to create because "CIS" means different things and raises different concerns at different organizations—academic centers are different from community hospitals, and inpatient care is different from outpatient care. Cost reduction as an overarching goal raises different issues than patient safety as the primary force driving adoption of computer-based provider order entry (CPOE) systems. A desire for high levels of decision support raises different issues than a desire for a basic system. Despite such variations, the experts at the study sites believed that certain themes were common across all CIS projects and could be addressed.

The following list of considerations reflects experts' consensus on key success factors for CIS implementation.

11.2 Consideration 1: Motivation for Implementing CPOE

The motivation for implementing CPOE influences where funding will come from, who will provide political support, and who will provide clinical leadership. It is important to consider what is motivating you and others in your organization to think about implementing CPOE.

11.2.1 Environment

- *Regulations:* Are local, regional, or national authorities contemplating requiring CPOE at a future date? For example, in early 2000, the state of California announced that hospitals and surgical clinics must have a plan for adopting technology to reduce medical errors by January 1, 2002, and must implement CPOE by January 1, 2005. More recently, a federal law governing the Medicaid program stated that by October 1, 2008, paper prescriptions would be reimbursed only if they are "executed on a tamper-resistant pad" (according to Section 7002(b) of the U.S. Troop Readiness, Veterans' Care, Katrina Recovery, and Iraq Accountability Appropriations Act of 2007). The intent of this law appears to be to reduce instances of unauthorized, improperly altered, and counterfeit prescriptions. Unfortunately, the added costs of buying new multi-tray printers or retro-fitting old printers, creating new security mechanisms to protect the new paper and the printers in which it will be used, reprogramming existing electronic medical record systems to recognize the new trays, and devoting additional time and energy to the implementation and maintenance of these new systems may reduce the likelihood that an organization can successfully implement CPOE.

- *Labor Shortages:* Will you have enough nursing and ancillary support personnel to staff clinical services? CPOE may be able to save clinician time through streamlined workflow processes, but it may be difficult to deal with the conflicting demands of the requirement for clinician input during the design, development, and implementation phases versus the need to maintain the existing levels of patient care. Following implementation of the system, there may be additional needs for nursing or ancillary providers to support the new clinical workflows.

- *Other Pressures:* Do you sense other environmental pressure? The Institute of Medicine report *To Err Is Human*, insurance company demands, the Leapfrog Group (which represents healthcare purchasers), and consumer demand all represent other external pressures faced by healthcare organizers as they attempt to deal with CPOE issues.

11.2.2 Workflow Issues

- *Administrative Needs:* Are administrators pressing for CPOE implementation? Administrative needs may include response to the previously described environmental pressures plus billing, quality assurance, and accreditation needs.

- *Clinical Needs:* Are clinicians pressing for CPOE implementation? Clinician needs may include the desire to apply information technology to improve patient care.

- *Efficiency Needs:* Is there pressure to improve efficiency? These needs may include lowering costs and/or increasing revenue.

11.3 Consideration 2: Foundations Needed Prior to Implementing CPOE

Successful CPOE implementations require effective leadership over extended time periods—in different forms and at multiple levels in the organization. Leadership is needed at the executive level to promote a shared vision and provide funding at the clinical level to ensure champions and buy-in and at the project management level to make practical, effective, and useful decisions. Before embarking on the serious undertaking of CPOE, organizations should determine whether the following conditions—which are necessary but not sufficient—are met.

11.3.1 Vision

Is there an overall shared vision for the organization regarding the purpose of CPOE (e.g., to improve patient care) as well as a common understanding of why the current state is suboptimal that would allow staff to embrace the concept of CPOE?

11.3.2 Leadership

Is there top-level leadership commitment that would commit to provide visible and unwavering support for CPOE? Institutions per se cannot commit; ultimately, it is people who commit. The organization needs a single, clearly identified CPOE project leader who holds a realistic attitude about what can and cannot be accomplished, coupled with the ability and knowledge to communicate the vision and articulate tangible objectives to educate administrators, and the interpersonal skills to foster teamwork.

11.3.3 Resources

Does the organization have adequate resources?

- *Infrastructure:* Is the technical network infrastructure appropriate?
- *People:* Is the clinical and project management staff available, knowledgeable, and ready?
- *Finances:* Does the organization have adequate finances to initiate and continue the project—coupled with real and visible commitment of the chief executive and financial officers?

11.3.4 Trust

Does the clinical staff trust and support the administrative staff? Conversely, does the leadership value, have faith in, and depend on the individual clinicians in the organization who will use CPOE?

11.3.5 Learning Organization

Is there a mental model throughout the organization (or one that can be created) that values constructive feedback, changes made for quality improvement, and continuous learning—kept in balance by leadership that can tell the difference between clinicians' requests for "what would be nice" versus "what is essential or critical for success?"

11.3.6 Sense of Urgency

Is there a compelling sense of urgency about implementing CPOE?

11.3.7 Vendor Readiness

- *Quality:* Are you satisfied with the quality of the product? Consider the quality and maturity of the product and service offered by the vendor. The stability and product quality of the vendor must be at least good, if not excellent. In addition, are sites similar to yours using the product as you plan to use it?

- *Stability:* Have you considered the long-term stability of the vendor in making your selection?

- *Relationship:* Can you put time and effort into forging a productive two-way relationship with the vendor? The vendor is your partner and helpmate in this endeavor—not simply a one-time supplier of services.

- *Innovation:* Is this vendor likely to offer more useful products over time than other vendors? Consider the speed with which the vendor is improving its product.

- *Flexibility:* How flexible is the vendor? Is the vendor able to integrate its systems with your existing suite of applications?

- *Reliability:* How reliable is the vendor in meeting deadlines and delivering high-quality code?

11.3.8 Maturity

Can the organization be considered mature and stable?

11.4 Consideration 3: Costs

11.4.1 Economic Aspects of the Decision to Implement CPOE

Financial considerations are of critical importance. Often, costs are underestimated because purchase of the software is only the beginning of financial outlays; other expenditures such as person-hours for training, back-fill for clinicians while others are in training, and post–"go live" support are more difficult to predict. Decision makers need to consider the following issues:

- *Timing:* Can you take a long-term view? Financial benefits may not be realized for a long time, and expenses over the short term may be significant. In addition, the organization should have the ability to commit additional funds quickly for good (unanticipated) causes (e.g., need to add more areas of wireless coverage). In the long term, the purpose of CPOE is to help patients.

- *Total Cost of Ownership:* Can you afford additional costs beyond those of hardware and software?

- *Productivity:* Can you afford a temporary loss of productivity? Consider that there will be a loss of productivity during training and the initial go-live period so that staff can take time to become comfortable with the new system; patient loads may need to be reduced or staffing increased during these times.

11.4.2 Dollars-and-Cents Considerations

- *Plan:* Do you have a good financial plan? Consider that by having a good financial plan, you will be ready to evaluate and address unexpected situations. The plan might take a broader view and look at other projects that may compete for money, time, and other resources.

- *Dedicated Funds:* Are there funds put aside for CPOE? Consider the level of adequate financial backing needed and assure that it remains dedicated to the project at hand.

11.5 Consideration 4: Integration/Workflow/Healthcare Processes

The manner in which a CPOE application alters and is integrated into existing environments and workflows is critical to its success. Users resent disruption of their patient care activities; thus implementers must consider a variety of issues.

First, time is of paramount importance to clinicians. Several facets of this issue must be carefully considered:

- *Response time:* Is it good enough? Consider how fast the system's response time should be for it to be tolerable for clinicians.

- *Ordering Time:* Is it time neutral? Weigh the tradeoffs so that the time spent entering an order is worth it—or at least time neutral—for the clinician. These tradeoffs might be easier access to information, ordering from multiple locations, and fewer calls about legibility.

- *Communication Time:* Will it be increased or decreased? Consider whether more (or less) total time will be spent gathering data and communicating using computerized CPOE rather than today's method.

Second, workflow issues need careful consideration:

- *Process:* Has the impact of CPOE on the work processes of physicians, nurses, pharmacists, ward clerks, laboratory personnel, registration personnel, and other hospital staff been carefully considered? CPOE needs to be seen as part of the employee's job; it must be integrated into the individual's workflow and that of the order communication process necessary for the execution of orders both in regular use and during CPOE downtime, and it must be used for all orders. In addition, clinicians must know how to view orders during construction, after entry, and after the order has been completed.

- *Changes in Work Structure:* Is it understood that work will change as a result of CPOE? CPOE may cause a redistribution of work and changes in the communication and decision making process. Users need to be able to visualize this change.

- *Strategy:* Is there an organization-wide change strategy? These workflow changes can be seen as a part of larger organization-wide change management strategies that have been tested under similar stresses of organizational change.

Third, integration must be planned carefully:

- *Scope:* Will all orders be done using CPOE? Consider the scope of CPOE to include the entire range of orders, not just one area.

- *Retrieval:* Is retrieval of information easy? Retrieval of other information such as medical records and medical literature needs to be integrated seamlessly into the workflow.

- *Embedding CPOE:* How well does CPOE fit with other hospital systems? CPOE needs to be integrated with other systems, such as the clinical laboratory system, pharmacy system, ADT/registration system, and other clinical systems, via interface engines and/or messaging protocols to create a complete electronic medical record system.

Fourth, readiness for integrating CPOE into the clinical workflow must be considered. Put simply, are the users ready for CPOE? Consider the level of physician readiness for CPOE and the communication and planning needed to increase readiness. At the same time, consider nursing readiness and staffing issues, plus the readiness of those who will receive orders generated by the CPOE system. For example, how will new, potentially life-saving orders be communicated reliably to nurses or others who need to be aware of them?

Other related considerations include the following issues:

- *Paper:* Have you decided how much paper can actually be eliminated in the process? Consider the role of paper and where in the ordering process its use might be tolerated.

- *Other Projects:* What are other high-priority projects? Consider other projects with which the CPOE project will compete, including both technology and other resource-intensive projects.

- *Fostering Use of CPOE:* Is there a plan for promoting usage? Will the project be mandatory for all medical staff? Hybrid computer–paper situations, for example, introduce frustrations and higher operating costs. Consider which mechanisms you might be able to put in place to facilitate and incent physicians to use the CPOE system.

11.6 Consideration 5: Value to Users/Decision-Support Systems

11.6.1 High-Level Decisions Related to Providing Value for Clinicians

Constituencies affected by CPOE implementation (e.g., physicians, nurses, ancillary department personnel) must understand the CPOE implementation "value proposition"; that is, they must do things differently but will receive some benefit in return. One benefit for clinicians is embedded CPOE decision-support logic that helps to improve patient care quality and/or to reduce costs. Related issues include the following:

- *Benefit:* How will clinicians benefit? The user must derive visible benefit in terms of improved workflow and a perception of "doing a better job" for the patient. The benefit of patient safety is a more intellectual and somewhat removed concept.

- *Results:* Have you implemented results review before undertaking CPOE? Experience with results reporting prepares users for CPOE systems.

- *Needs Analysis:* Have you done a needs analysis? Analyze user needs carefully; do not just give people what they ask for.

- *Communication:* Have you considered the impact of CPOE on communication flow? Consider that adoption of a CPOE system tends to decrease face-to-face communication.

- *Involvement:* Is there a plan for involving physicians? Physician involvement is needed from the start and throughout the CPOE development process.

11.6.2 Implementation of Decision Support

Being able to provide decision support is an important benefit of CPOE. The following issues need consideration:

- *Content Determination and Maintenance:* Is there a plan for ongoing decision making about decision-support content? Consider putting a process in place to determine which kind of decision support should be implemented and how to oversee and maintain it.

- *Efficiency:* Will decision support improve efficiency? Consider how to implement decision support in such a way that physician efficiency can be improved. Examples include the use of order sets and therapies based on diagnoses.

- *Alerts:* Will you provide alerts? CPOE systems can deliver value even when they do not provide alerts and reminders. Decision-support capabilities can help the clinician make faster and more confident decisions; offering constrained choices is a form of decision support. Alerts, by contrast, can become noise and aggravate users. Drug–drug interactions and drug allergies alerts are useful and fairly easy to obtain.

- *Readiness:* Are users ready for decision support linked to CPOE? Consider decision support when assessing readiness for CPOE.

11.6.3 Other Considerations Concerning Value to Users

- *Perception of Efficiency:* What value does the CPOE system have for the individual physician? There needs to be a clinician perception that the software makes the physician more efficient. Clinical users must be shown that CPOE usage is not clerical work, emphasizing what cannot be done via manual, paper systems.

- *Technology:* Is the proposed technology far enough advanced?

- *Benefit:* Are benefits for clinicians easy to see and describe? Demonstrable benefits are needed to assure continued, sustained use. Users must participate in development of decision-support features that affect them, and be adequately trained in those features' usage.

- *Education:* Is there a plan for educating as well as training users? Educate users about the limitations of CPOE as well as its capabilities. Users must be trained before implementation and on an ongoing basis thereafter as CPOE systems evolve. Users must understand where the CPOE system provides "help" and where it may not, and CPOE behavior must be consistent (e.g., for drug–drug interaction, drug–allergy interaction, duplicate medications, duplicate labs, expensive tests, suggested drug level monitoring).

- *Patient Care:* Are there clear benefits for patient care? Emphasize that CPOE is intended for the good of the patient, not as a means to reduce the bill.

- *Order Sets:* Is there a plan to implement order sets—that is, groups of orders to manage a specific disease state or a procedure (pre-bronchoscopy, post-bronchoscopy)? These order sets must be developed, reviewed, and maintained for clinicians' personal and/or departmental usage. If available, they can provide local control as well as perceived benefit.

- *Intentions:* Is the system designed so that the clinician easily understands the status of an order? Make sure the system does what the user intended.

11.7 Consideration 6: Vision/Leadership/People

Effective leadership is needed at several levels in the organization: at the executive level to get funding, at the clinical level to attract champions and gain buy-in, and at the project manager level to make practical, effective, and useful decisions. Both the leadership and software need to be flexible enough to be able to make modifications that address identified concerns and problems. These leaders may or may not be the same person. A shared vision needs to underscore work at all levels.

11.7.1 A Shared Vision

Is there a shared vision regarding the purpose of CPOE to improve patient care, and are there stated goals for fulfilling that purpose? Do physicians regularly play a role in strategic planning and IT decisions?

In terms of communication, are there physician leaders and champions who can effectively communicate the shared vision? Is there an ability at all levels to communicate the vision and articulate tangible objectives?

Are there enough people who feel that the current state is intolerable and that change is needed?

11.7.2 Considerations at the Highest Level in the Organization

- *Commitment:* Is there real and visible financial and administrative commitment by leadership at the chief executive officer level?

- *Persistence:* Does the leadership exhibit persistence in striving toward ultimate project goals?

- *Trust:* Is there a sense of trust, credibility, and communication between the administration, implementation team, and clinician users?

- *Strength:* Is the leader someone who can make a decision on his or her own if strategies for reaching consensus fail?

- *Function:* Can the leader differentiate between the CPOE functionalities that clinicians want and those that they actually need for patient care?

- *Urgency:* Does the leader sense a level of urgency, from either external or internal motivators, about implementing CPOE such that it is a top priority?

- *Style:* Does top leadership understand its own leadership style? Leadership does not need to be charismatic, of course: Different leadership styles can be equally effective.

- *Perceived Value of Clinicians:* Does leadership have faith in, value, and depend on individual clinicians in the organization to make implementation succeed?

11.7.3 Considerations at the Clinical Leadership Level

- *Ability:* How well do leadership skills fit different phases of implementation? The clinical information technology leadership must have the ability to use the management style appropriate for needs at different stages of the project:

  1. Pre-implementation: Develop a vision, get funding, and identify individuals who will be key for the implementation, elicit involvement from these key people, and exhibit other strategic and tactical planning skills.

  2. Implementation: Hire staff, deploy staff where and when they are most needed, keep up the spirit of the staff doing the work, and use other communication, publicity, and personnel management skills.

  3. Post-implementation: Establish the maintenance phase, create an environment for ongoing system improvement, and provide management systems for the long term.

- *Attributes:* How well do leadership attributes fit the task? This leader must have clinical credibility (be respected by physician peers) and must be persistent, consistent, accountable, and thick-skinned.

- *Realism:* How realistic a view does the leader have? This leader must maintain an organizational anticipation and excitement for the project without overselling it and creating unrealistic expectations.

- *Educator:* Can the leader educate administrators? The leader must be good at educating executives and keeping them up to speed.

- *Feedback:* Will leaders listen to constructive feedback? The decision makers need to come to user feedback sessions. They need to actively solicit both negative and positive feedback, and respond to that feedback in a timely, demonstrable fashion. Identified problems must be addressed expeditiously.

- *Golden Rule:* Leadership should follow the golden rule—do unto others as the leader would have done to himself or herself.

- *Teamwork:* Does the leader foster teamwork? The person should be able to form a great team, inspire clinician involvement, and be viewed as an advocate by clinicians.

11.7.4 Clinician- or Physician-Level Champions and Project Leaders

- *Clinical Skills:* Are these leaders clinically trained? At least one leader at this level needs to be a clinical person, albeit not necessarily a physician. This position should be filled by a paid person with visibility who is at least partially excused from competing clinical duties.

- *Involvement:* How heavily involved in implementation should the clinical champions be? The clinician leader(s) may not be at the top of the project leadership hierarchy, but a clinician must be involved at a visible and influential level.

- *Technical Knowledge:* What is the leader's level of technical knowledge? At least one of the clinical project leaders must have enough technical knowledge to be able to challenge technical staff and vendors.

- *Sympathy:* Is the leader sympathetic? Leaders need to be charismatic and clinically credible, although not necessarily technically trained. It is most important that these individuals be seen as understanding of and sympathetic to the needs of clinicians.

- *Opinion Leaders:* Are there other opinion leaders identified? The effective clinician leader will enlist the assistance of both senior credible champions and people perceived to be technical opinion leaders.

- *Role:* Is there one identifiable top leader in this group? There must be only one clinical project leader, and the roles and duties of the clinician leader must be clearly delineated.

11.8 Consideration 7: Technical Considerations

Technical details to consider as part of a CPOE implementation include strategic considerations, user considerations, task completion flexibility, and the quality of the application—from customizability to user friendliness.

11.8.1 Strategic-Level Considerations

- *Security:* Is there a security plan? Data backup and disaster recovery are significant considerations for mid-level managers and higher. While downtime may be necessary for data backup, it should be minimal to cause the least disruption to clinical workflow.

- *Customization:* How customizable is the system? Consideration needs to be given to the amount and level of customization allowed by a particular CPOE system, including the ability to provide decision support where needed. It should be customizable by an on-site analyst to accommodate variations in workflow and procedures from department to department, from unit to unit, and from shift to shift.

- *Replacement:* Are there any special considerations for replacing older systems? Even sophisticated users may have difficulty adjusting to a new system that does not meet workflow needs as well as the older, already customized system.

- *Data:* Is there assurance of high-level data quality? Accurate and reliable data must be maintained at the highest possible level to ensure clinician acceptance.

- *Connections:* Can the CPOE system interface with existing and planned systems? Interfacing capability among systems from different vendors is important. In particular, interface engines or hubs and HL7 protocols need consideration.

- *Access:* Has a risk analysis been done? Security of access and confidentiality issues must be seriously considered, especially given the need to comply with HIPAA regulations. Is there is a plan to authorize all users who need access to the system (e.g., attending and house staff physicians, nurses, medical assistants, and unit secretaries). Advantages and disadvantages of a single sign-on need consideration as well.

- *Remote Access:* Is there a need for access from remote locations such as the patient's home, nursing homes, or other sites? If this kind of access is needed, does the vendor support this capability? How will you manage the remote desktops to ensure they are "safe" (e.g., free of viruses, adequately secured, contain no personally identifiable data) to be on your network?

- *Infrastructure:* Is the network infrastructure stable?

11.8.2 User Considerations

- *Escapes:* Are there escape routes for frustrated users? Consider establishing escape mechanisms for nonstandard, unusual, and complex orders. For example, the interface might include a section in which plain free-text typing is permitted.

- *Interface:* How easy is it to use the interface? There needs to be a consistent user interface that is intuitive, easy to navigate, and efficient. It should include a logical flow from one screen to the next.

- *Time:* How time-consuming is the system from the user's point of view? A new CPOE system will likely increase a clinician's charting time or—at best—be time neutral. At the very least, the system needs to be secure, fast, and reliable.

- *Clerical Tasks:* Will users view use of the CPOE system as clerical work? Consideration should be given to how the rationale for a new system is communicated so that clinicians do not perceive it as clerical work.

11.8.3 Flexibility in Task Completion

- *Style:* Can the system fit different work styles? Because individual work styles differ, consider allowing multiple ways to do the same thing. For example, keyboard equivalents for mouse actions help accommodate differing work styles.

- *Customization:* Can users customize some things themselves? System administrators must strike an appropriate balance between customization and standardization. In addition to allowing systems analysts to modify the system, there should be options for users to customize screens as well.

- *Decision Support:* Has the addition of decision support been carefully considered? Carefully consider the provision of decision support to avoid overloading the clinician with messages. One approach to this feature is to allow tuning of a drug interaction alert to account for the severity of the interaction.

11.9 Consideration 8: Management of Project or Program/Strategies/Processes from Concept to Implementation

11.9.1 Highest-Level Considerations

- *Impact:* Have you carefully considered the impact on workflow? Too narrow a concept of implementation can derail the project. Realize that reengineering the order-entry process will affect other clinical and ancillary processes.

- *Strategy:* Is there an overall strategy for improving care? CPOE needs to be part of a larger strategy to improve patient care.

- *Management:* Have the people issues been carefully considered? Sound project management during implementation and ongoing program management post-implementation must be planned with human factors in mind.

- *Scope:* Is there a defined scope to the project? An emphasis on sound project management is essential.

- *Treatment of Others:* Do you adhere to the golden rule?

- *Detail:* Are plans detailed enough but not overly so? Perfection is the enemy of the good; do not allow exaggerated attention to details to jeopardize the overall implementation goal. Keep it simple; strive for excellence, not perfection.

- *Goals:* Are there clear and measurable goals?

- *Communication:* Is there a plan for constant communication with users and implementation staff?

- *Expectations:* Are expectations reasonable and achievable, yet maintain excitement for the project?

- *Relationships:* Have you anticipated significant changes in clinical relationships and planned accordingly?

- *Ambition:* Have you carefully considered how ambitious your goals can realistically be? Without disciplined project management, outside influences may force a project implementation pace that is too ambitious.

- *Consensus:* Can you balance consensus with directive leadership? Too much emphasis on consensus will slow project implementation; too directive an approach can decrease user involvement.

- *Downtime:* Have the implications of downtime been considered and procedures to manage them been established?

11.9.2 Mid-Level Considerations

- *Consultant Expertise:* How will consultants be used? Weigh the need for internal expertise versus consultants; avoid becoming too dependent on outside consultants by developing internal expertise.

- *Critical Mass:* Will you be ready for the important moment? Recognize that a critical window of opportunity will arrive where a critical mass of users is using the system. It is vital that the implementation team capitalize on this milestone.

- *Long-Term View:* Can leaders adopt a long-term view? The administration and information technology team need to take a long-term view of CPOE implementation.

- *Early Objectives:* Have you identified early wins? Categorize implementation objectives as easy, hard, or hardest to implement; start where early success is expected. Consider gaining an early victory by making results reporting available early in the implementation process.

- *Vendor:* How carefully have you chosen a vendor? Exercise care and choose an experienced vendor. Realize that the implementation's success may depend on cooperation, if not synergy, with the vendor. Expect and depend on a long-term relationship with the people in the vendor organization.

- *Clinicians:* Is there a plan for involving clinicians? Allow interested (and encourage uninterested) clinicians from all specialties to participate in product selection and local customization efforts. Turn ardent opponents into ardent supporters; convince the skeptics and curmudgeons.

- *Users:* Have you considered all clinical users? Carefully consider that the "P" in CPOE can stand for "professional," "physician," "provider," or "practitioner." Realize that the CPOE function will have effects on organizational workflow beyond order entry alone and that other functions, such as results reporting, will be affected as well.

11.9.3 Lower-Level Considerations

- *Workarounds:* Are workarounds available? Provide simple workarounds for occasional users, such as a text entry option. At the same time, provide some mechanism by which all users may improve their skills so that dependency on the workaround does not develop.

- *People:* Are variations in people's ways of doing things being considered? People engineering needs to be concurrent with software engineering needs; realize that workflow redesign and attitude maintenance may be required with CPOE implementation.

- *Metrics:* How good are your metrics? To do good project management, you must have specific metrics; it is vital to know what is working and which elements must be improved. Develop "before" as well as "after" metrics.

11.9.4 Roles

- *Accountability:* Who is accountable for what? Assign and expect personal accountability for all project tasks.

- *Clinicians:* Is there clinical involvement at the leadership level? There must be a clinician representation within the project leadership. Recognize that providers play a pivotal role in the implementation's ultimate success.

- *Champions:* Have champions been identified? Clinician champions should be identified early and supported and relieved of some other duties so that they can fulfill this role.

- *Leaders:* Is there an identified clinical leader? There is need for a strong leader with a foot in both the clinical and technical camps, possibly someone with medical informatics training.

11.9.5 Localization

- *Modification:* Can you modify the system on site? You need to be able to do some modification at the institutional level.

- *Customization:* Can users customize some things themselves? For a fast win, allow more individual customization.

- *Balance:* Have you considered the balance between customization and standardization? Consider how local modifications might affect vendor upgrade paths. Too many local modifications may make it difficult to implement the vendor upgrade, as the upgrade may not support the modifications. Even if it does support the modifications, each modification will slow down the ability to upgrade.

11.10 Consideration 9: Training/Support/Help at the Elbow

Considerations for training and support include the concept of "help at the elbow." In addition to the symbolic importance of supporting the users by being present when they initially work with the application, intensive support at "go live" time allows the implementation team to have direct experience with what is and what is not working well. Most successful implementations have had more post–"go live" support than pre–"go live" training. In fact, most sites have made support available on a 24 × 7 basis for at least several days following the go-live date.

11.10.1 Support

- *Help:* Is there a plan to provide help at the elbow? Skilled support staff should be available all the time during the CPOE system's implementation and much of the time post-implementation; err on the side of too much user support.

- *Training:* Is there a training plan for support staff? The ideal support staff members will be high-quality, patient, thick-skinned support people who have good people and communication skills and can teach others to use the application.

- *Translators:* Can support staff act as translators? Support staff should be able to translate between the clinical and technical realms.

- *Online Help:* Are there provisions for online help? Also provide different mechanisms for help such as online help in addition to on-site help.

- *Help Desk:* Does the help desk operate on a 24 × 7 basis? Is it staffed by experienced technical support personnel? Are the processes and tools mature?

11.10.2 Training Methods

- *Training:* Will users train other users? During implementation, consider using successful users to train the next set of users.

- *Tools:* Will multiple training methods be used? Provide multiple learning tools and methods, including computer-based training.

- *Plan:* Is there an initial plan? Plan to provide sufficient initial training, and err on the side of too much training.

- *Updates:* Will there be updates? Consider providing ongoing updates that also go back and reiterate prior information.

- *Monitoring:* Is there a provision for monitoring proficiency? Consider adoption of a continuous retraining program to make sure that physicians are using systems effectively.

11.11 Consideration 10: Learning/Evaluation/Improvement

CPOE implementation is an ongoing effort that will benefit from continuous improvement. For this reason, it is important that mechanisms for feedback and modification of the system be in place.

11.11.1 Higher-Level Considerations

- *Problems:* How will problems be addressed? Consider carefully planning a process for problem identification and problem resolution involving the users. Make sure there is a process for responding to problems in a timely manner.

- *Feedback:* What is the formal evaluation plan? As formal evaluation takes place, share the results of the evaluation with the users so that they can gauge responsiveness.

- *Testing:* How will you test the system? Think about how you test whether the system is good enough to go live. Carefully consider how the system will be "test piloted" without putting patients, or the organization, at risk.
- *Continuous Improvement:* How will you continuously improve the system? Understand that you are never done—continuous improvement is essential.
- *Learning:* How will your organization learn from both its successes and its mistakes? Be a learning organization—learn from evaluations.
- *Revisiting:* Is there a plan to revisit strategic decisions on a regular basis?
- *Continued Training:* Is there provision for regular training sessions (e.g., brown bag lunches)?

11.11.2 Strategies

- *Response:* Is there a process for responding to problems? Make sure the organization is capable of providing a quick response in case system flaws appear.
- *Escapes:* Is there an escape mechanism? Provide an escape mechanism such as free-text entry; it can be a great source of feedback, allowing you to see how to improve the system.
- *Testing:* How will the system be tested? Have moonlighting house staff test the system. Have adequate integrated testing.
- *Pilot Tests:* How will pilot tests be conducted? Pilot software in small groups and improve it if necessary before rollout to the larger organization.
- *Mentoring:* Is there a mentoring system? Establish a buddy or mentoring system so that clinicians can share their expertise in developing order sets and templates and exchange tips.

11.12 Final Thoughts

Each consideration should be reviewed by the leadership and implementation team of any organization considering CPOE installation. Some issues will be more easily addressed than others; some will be more relevant to one particular organization than others; and some will be more applicable at different stages in the implementation than others. Some of these questions and issues will have clear and obvious answers, but most will not and will require effort to address. Organizational representatives should focus on the difficult-to-answer questions rather than avoiding them. All of the detailed considerations listed in this chapter are relevant to a successful implementation, and failure to address any of them could put the success of the CPOE project in jeopardy.

The importance of strong executive leadership at the highest levels in the organization in a CPOE initiative should not be underestimated. Leadership is a thread that runs through many of the major considerations. Administrative leaders, acting on behalf of the organization, must believe viscerally that adoption of CPOE is in the best interest of the institution—and they must be able to communicate that feeling throughout the organization.

Clinical leadership must also be committed to CPOE and communicate this commitment to the clinical staff, who will typically be less than excited about the prospect of CPOE (due to natural resistance to change). Clinical staff will have real concerns that the time needed to complete work will increase once such a system is implemented. Arguments about improved safety may appear to end users as vague and intangible. Given end users' natural skepticism, clinical leaders must work strenuously to communicate (and physically demonstrate) to their staffs how CPOE will provide opportunities for improved quality and efficiency. Administrative and clinical leaders must work together to create a strong sense of "common will" to overcome obstacles that will be encountered during a CPOE implementation.

12 The Importance of Special People

Based on: Ash JS, Stavri PZ, Dykstra R, Fournier L. Implementing computerized physician order entry: the importance of special people. *Int J Med Inform.* 2003;69(2–3):235–250.

12.1 Key Points

- There is a wide variety of special people involved in the implementation of computerized physician order entry (CPOE), including administrative leaders, clinical leaders (champions, opinion leaders, and curmudgeons), and bridgers or support staff who interact directly with users.

- The recognition and nurturing of special people should be among the highest priorities of those implementing CPOE.

12.2 Introduction

Diana Forsythe's classic paper, "New Bottles, Old Wine: Hidden Cultural Assumptions in a Computerized Explanation System for Migraine Sufferers,"[1] describes "the problem of user acceptance" so common in clinical informatics applications. It points out that we often blame the users for not embracing new systems, yet a system may "embody perspectives that may not be meaningful to or appropriate for their intended users".[1]

In our previous work,[2] we identified three assumptions concerning computerized physician order entry (CPOE):

1. Order entry is a linear process, beginning with the physician entering an order and ending with its being carried out.

2. Physicians are recalcitrant: They resist behavior change.

3. Structured input of data for orders is good for two major reasons: (1) for legibility and (2) for analysis purposes.

These assumptions are the "old wine" put into the "new bottle"—that is, CPOE.

Diana Forsythe[1] asked, "Whose assumptions and whose point of view are inscribed in the design of technical systems? Who will benefit from adoption of a given system, and who stands to lose?" In our study, we discovered that CPOE is not a linear process; the order communication process is, in fact, exceedingly nonlinear, with many people involved in formulating the idea of the order, modifying it throughout the process, and carrying it out and documenting it. We found that physicians are not necessarily recalcitrant: They simply do not wish to spend additional time during the ordering process—they would prefer to spend that time with patients. Finally, structured input benefits administration and it fits the capabilities of computer systems, but it does not necessarily benefit the physicians who prefer entering or writing free text.

The Forsythe paper[1] ends with the question, "Who will monitor the hidden cultural assumptions built into computerized tools for medicine?" That is the question addressed in this chapter.

12.3 Special People

When asked, "What made your implementation successful?" many interviewees who participated in our study answered that a certain individual was key and went on to describe that individual's attributes and actions in detail. Special people were high-level leaders—either nonphysician clinicians who assisted with the implementation or physicians who played a special role during implementation. Their roles spanned disciplines from administration to information technology to the clinical realm. Because they lived in more than one world and knew the vocabulary of each realm, they could interpret from one to the other. In addition to vocabulary, they could interpret disciplinary culture.

Figure 12-1 depicts how these special people have overlapping, intersecting roles, sometimes holding an administrative role or a paid position in informatics, and

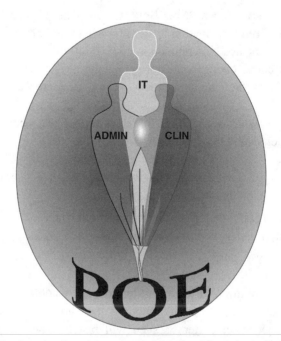

Figure 12-1 The overlapping roles of special people: administrative, information technology, and clinical staff.

sometimes practicing another clinical specialty such as pharmacy while providing technology support. Each has a view of CPOE that is colored by his or her role or combination of roles. We divided the roles into three major levels:

- The leadership level, including the chief executive officer (CEO), chief information officer (CIO), and chief medical information officer (CMIO).
- The clinician level, including champions, opinion leaders, curmudgeons, and the clinical advisory committees.
- The bridger/support level, which includes those individuals who do training and support and interact directly with users. We call these people "bridgers" because they translate user needs to the higher levels and vice versa.

How are these special people alike? Their behavior as interpreters between the technology and clinical worlds is key. Even the CEOs who were at these successful sites during implementation of CPOE exhibited the ability to understand the clinicians and their behavior. All held a vision of the future and the way CPOE fits into the electronic medical record. Many personal attributes were universal, including stability through adversity, steadfastness, willingness to take the initiative, and toughness.

12.4 Administrative Leadership Level

The administrative leadership level includes the CEO, CIO, and CMIO. The CEO administers the entire organization and information technology and clinical leadership report to him or her. The CIO generally heads the information technology unit and often the information services unit; he or she reports to the CEO. The CMIO is also an administrator but has clinical information as a focus and reports to one of the former levels and/or a chief medical officer (CMO).

At this level, the roles of administrators and clinicians often overlap, as Figure 12-1 indicates. For example, the CEO or the CIO may have a clinical background; the CIO may be part of the administrative team and at the same time head IT. Leaders also include individuals who serve in a CMIO role: This person usually plays roles in administration and IT, has a clinical background, and may even continue to practice medicine. These people may hold other titles as well, but the role is one that is somewhere between the CEO and the users of CPOE.

12.4.1 The CEO

Leadership and top-level support were usually cited among the most important success factors for implementing CPOE. We define the CEO as the individual who heads the entire organization, even if the person has a different but equivalent title. This person is depicted in Figure 12-1 as filling the "Admin" (administration) role.

At sites that had successfully implemented CPOE systems, CEOs exhibited a number of common characteristics. First, they provided top-level support and vision:

What made a huge difference for us is that we had the unwavering backing of our director [CEO], and then the people who were on the next level [CMIO] also [said] this was going to be done.

Of course, financial support is key. One CEO said:

Hire really good people, give them the tools, and stay out of their way. I mean, give them support . . . We had to put a lot of money into the infrastructure. We had the design done, and then I just devoted the dollars that it would take. I took equipment dollars, the construction, the nonrecurring maintenance dollars, and, without hampering patient care, devoted the lion's share of those for a couple of years into that.

Second, the successful CEO holds steadfast when it comes to CPOE. One informant said:

What kept us going here during the tough times is the fact that our administration . . . said [to IT], "You know, guys, do whatever you can." They realized the importance of keeping it going: "Just tell how much it's going to cost, and do what you have to do."

Third, the effective CEO connects with the staff:

- "He was here over 25 years and politically astute and more of a team builder and got people working together."
- "I was struck by how much X is like the legend of Ed Hawkins at El Camino. People seem to see him in the same way. He also goes out on to the floors and considers himself one of them." (R. Edwin Hawkins was El Camino's first hospital administrator and went on to lead the hospital for 25 years.)
- "He actually knew most of the employees by their first names, and he was very genuine. He would come around to the departments, and he was very open. You could go up and see him."

Fourth, the CEO listens: "They listened to nursing carefully in developing the system." At sites that had successful CPOE rollouts, participation—not buy-in—was solicited. A bidirectional flow of communication was deemed important, and these CEOs really listened and accepted help from users in planning for and implementing the system: "You get louder because you think people aren't listening."

Fifth, these CEOs acted as champions. A "champion" is defined here as someone who fights for the cause—for example, implementing CPOE. As one CEO said, "You really have to be a cheerleader and an advocate as well as a decision maker."

12.4.2 The CIO

The CIO often reports to the CEO and is usually directly responsible for CPOE implementation. Our informants told us that this is a position that requires savvy political skills and a thick skin.

Among the characteristics shared by CIOs at sites that had successfully implemented CPOE was the ability to select champions. Successful CIOs know how to select clinical leaders:

> X was a master at this—finding the leaders and giving them a lot of special attention, helping them understand how [CPOE] could help them, like with personal order sets. Then this leader [would] say, "This saves me time and it makes the whole process more efficient"—and it saves me getting a phone call.

CIOs also won support from other employees:

> He did have a lot of political moxie and he was very determined. He grew his department; it got bigger and bigger.

In addition, CIOs tended to possess vision, just like the CEO:

> He was pretty goal-oriented. He would see a goal and work toward it, regardless of what was in the way.

Finally, the CIO maintained a thick skin. As the person directly responsible for CPOE, the CIO may also get the blame for shortcomings:

> X was embattled much of the time. [That] was my impression; I don't know if he'd say that because he was constantly in the role of defending it.

12.4.3 The CMIO

The CMIO role was called different things at different hospitals in our study, but each facility nevertheless had one or two people filling this role. These physicians straddle the line between medicine and administration (see Figure 12-1, where they are labeled "Clin" for "Clinical"). They have many of the same attributes as the CEO and CIO.

First, CMIOs have an uncanny ability to interpret information. They understand the business of the hospital and try to translate it for their clinical colleagues, and they attempt to explain the culture of medicine to administrators.

Also, like their counterparts the CEO and the CIO, the CMIO possesses vision:

- "We were very fortunate to have Dr. X, who was able to hold the bigger vision of this project, because I'd seen other facilities that haven't had that luxury or didn't hire ten [clinical support staff] to bring it up."
- "I think, for the leaders, you need to have a firm vision and a team that can help you get there, and to stay the course."

Like the CIO, the CMIO must have a thick skin:

- "Make the corrections that are necessary, . . . part of being a leader is also taking the input that you get, but not necessarily abandoning [the project] when things get tough."
- "Dr. X was a much hated man, but a nice guy."

The CMIOs at sites that had successfully implemented CPOE systems also influenced their peers. One support staff member who had trouble dealing with a physician said the following about the CMIO:

> It just became clear that I wasn't going to be enough . . . so I had [the CMIO] work with her [the physician], initially. And, given his very calming personality and the fact that he was relating to her physician-to-physician, that helped a lot.

A CEO said of his CMIO:

> He has the clinical relevance to the others because he works beside them. That was the kind of person I needed—somebody who had credibility and good strong skills and a real fire to make an electronic record work for us.

While the CMIO needs the support of the CEO and the CIO, he or she likewise needs to support the critical support staff members who deal directly with the users.

> They [the CMIO and those at higher levels of the administrative hierarchy] kept telling us that we were important, and whenever we needed to have a work–life balance, they were supportive of that because they wanted us to feel valued.

As one user also noted, "One part of that [CMIO] role is being a champion."

12.5 Clinical Leadership Level

Successful sites put resources into identifying and sometimes even hiring physicians and other healthcare professionals to socialize the idea of CPOE throughout the organization. We identified four types of leaders in this category.

12.5.1 Champions

We define champions as individuals who fight for the cause, who believe in CPOE and its importance as part of the EMR, and in the value of technology for achieving goals.[3] The champion does not necessarily need to be someone with advanced technical or clinical skills. In addition, champions are sometimes looked on skeptically by the majority of the workforce because they are early adopters and perhaps have such a passion that their view does not appear to be balanced. For example, about one champion, it was said in field notes that it was "hard to get him to talk about weaknesses [of CPOE]."

Nevertheless, the need for a champion to drive CPOE forward is clear. As one leader said, "Certainly, I would want to have some clinical champions identified before taking on anything of this size, any project." One CEO, when asked how he would do it differently next time, replied, "I would get the clinical champions in place earlier." Other comments were equally revealing regarding the need for a champion:

- "I think you need a number of clinical champions within the areas, the clinical areas, separate from the clinical informatics group."

- "[It is very] important . . . at the outset to deal with people who would buy in or are going to have the insight that this is a change for the better—that once you [implement CPOE], you're not going to regret it; in fact, you're going to love it. If you identify these people—we called them champions, and I was one of them—they become the agents who are going to sell this change among their peers."

The champion holds steadfast to the notion that the change is for the better. One informant told of a clinical champion who failed because he had an interest but not a "fire in the belly for technology." To succeed, the champion needs to be persistent:

The head of the physician group down there . . . was very positive. He's fairly soft-spoken but he doesn't go away; he just keeps saying, "We need to do this."

The champion also influences his or her peers. At the sites participating in our study, sometimes the champions were recruited to help directly with training their peers on a one-on-one basis:

If you were unsure of what to do, a doctor would sit next to you and you weren't made to feel that you were incompetent. There was a lot of one-on-one.

Perhaps not surprisingly, the physician champion understands other physicians. One of the champions expressed the feelings of his peers about concerns with response time following adoption of a CPOE system:

Did somebody talk to you about response time, how critical it is? In my experience, that's the key in getting doctors to use anything, because we're all Type A [personalities] . . . there were slowdowns and it was terrible, but doctors are very impatient, . . . they'll say, "The hell with this thing."

12.5.2 Opinion Leaders

An opinion leader is an influential physician who may be in favor of the new system or work against it. Opinion leaders are respected by peers, usually for both professional and social skills.[4]

Opinion leaders provide a balanced view:

They were just people with common sense, I guess. You know there're a lot of different personalities, and if I were to try to pigeonhole them, I would consider them for the most part to be more level-headed, even-tempered.

These individuals also influence their peers in a positive way, as the following comments suggest:

- "Using peers to influence peers to go through the change, rather than giving a directive [that] everybody should do it, [is a good idea]."

- "Dr. X worked closely with [the CIO] . . . he liked computers, but he kept his practice and did this also. He had a large contribution, not to the initial buy-in but later. He was very active; he was a surgeon. A lot of activity happens around surgery lounges."

- "The hospital paid him a salary. He was interested, [serving as a] quality assurance guy for a while. He had that ability to lead and get doctors together, but he had help from the hospital to do paperwork, etc. [He was] very respected on staff. He was a neutral guy, kind of in the middle."

12.5.3 Curmudgeons

We define a curmudgeon as a skeptic who is usually quite vocal in his or her disdain for the system. Our informants felt that convincing the curmudgeon is a key to system implementation success because the loudest skeptics may turn into the staunchest supporters.

Curmudgeons, for example, may provide highly useful feedback:

I mean, the person who screamed the loudest was an X fellow. She was a nightmare, frankly—but she also made the biggest changes and helped us learn what we needed to do to make the order sets work.

The curmudgeon also furnishes leadership:

You can become a physician leader by being an irritant, screaming and yelling every time anything happens; you get acknowledged and catered to. You can be a skilled practitioner who's respected by peers and be charismatic. You can also have political skills where the others who don't [have the same skills] will come to you . . . they [the CMIO and support staff] targeted them [the curmudgeons].

12.5.4 The Clinical Systems Committees

This category includes the committee of physicians and other clinicians charged with oversight of CPOE. At the sites included in our study, the work of this hospital committee of special people was often described as an important success factor for CPOE adoption by our informants. This committee may have different names at different hospitals, but it is generally an advisory committee that includes clinician members focused on CPOE alone or on clinical systems in general. At each of these successful sites, this committee was given serious responsibility for making recommendations to be followed by the leadership.

The committee serves an important problem-solving role:

- "If there is a problem, then it's up to the [clinical systems committee] to deal with it."

- "Probably 20 people got involved to try to work toward solving the problem."

Another role the committee plays is that of fostering teamwork. Generally, different professions are represented in its membership, including nursing, dietary, medical records, and others. As one informant said about this multidisciplinary committee, "We'd get them together in a room . . . where decisions were made." A committee member said the following about a CIO:

> He wanted a group of clinicians to serve as an oversight. He didn't want [the committee members] to be senior management, because he knew [senior managers] rarely practice, so they sort of miss something. They don't have a lot of credibility in the trenches, unfortunately, and they also don't understand a lot of the issues. So we got a bunch of people sort of at my level in their careers who are still practicing and spend a lot of time on the wards.

An information systems committee member and an opinion leader at a different site said:

> I am on that committee, and [the] other [committee I am on] is the organizational committee, which is much more of a business type of approach. I basically sit and listen or represent the viewpoint of perhaps a fair number of physicians who have moderate [computer skills] . . . a modest functional use of any type of computer.

12.6 Support Staff Level: Bridgers

At each successful CPOE-implementation site, there was a cadre of special people—clinicians bridging the gap between the information technology group and other clinicians—known as "bridgers" or "translators." One site called them "clinical application coordinators." Working at a grassroots level, these individuals do more than champion the system: They work for results. These are the people who understand the work, are paid to train and assist physicians in using CPOE, and make changes in the system. Although their personalities differ, they share a respect for physicians, a belief in the benefits of CPOE, and a value system in which the patients come first and the users next. Most often, those who train and support CPOE are nonphysician clinicians. Informants in this category came from nursing, pharmacy, and physical therapy backgrounds. These are the people who "bend over backward" to help, and in so doing make the difference between success and failure in rolling out a CPOE system.

12.6.1 Need for Bridgers

Bridgers are clearly needed to ensure the success of CPOE. As one CEO said, "We made a decision that we needed to get enough resources into the support activities and the training activities to make this a success." According to another interviewee:

> It made no sense to put the organization through the pain of going from a paper record to an electronic record if we didn't have the support in place to make it a success. I had seen that when we even wanted to take a baby step and hadn't put the resources there, it failed.

All of the successful sites had help available during implementation all day, every day—that is, "help at the elbow." All continue to have valued assistance easily available in the post-implementation environment as well:

- "[Betty] was the key person to help with—she went to other units to facilitate implementation. I think [Betty] was the difference."
- "We had to track [use of the CPOE system]. We knew exactly who these people were that weren't using the system. We sat down and helped them make their own personal order sets; we just sat down and held their hand."

These were also people who were known in the organization:

> Another thing that helped with implementing the system was that X and her staff were here so long and already respected. When they went out selling this thing, when people had problems and called, they knew who they were talking to.

12.6.2 Bridgers' Role in Making Changes

Bridgers generally interact not only with the users, but also with the system vendor:

> The vendor did all the coding; our job was to describe problems to the vendor and work with them and beat them over the head. Gradually we took over our own coding.

There are always tweaks that need to be made, even at the site with the world's longest CPOE use:

> We have a very good staff; they're very technical. We do routine changes every week—we try to do it on Wednesdays—and we use the committees that we have for approvals.

Changes are suggested only if users feel free to request them:

> If you get your users involved, they won't be hesitant to call you and say, "Do you think this would be a good idea?" or, if there's a problem, "Fix it." ... the user out there who uses [the CPOE system] all the time says, "Well, why didn't you kind of do it a little bit differently?" We strongly encourage people to tell us [when changes are needed].

12.6.3 Bridgers' Role in Providing Training

Bridgers are also the people responsible for most training in use of the CPOE system, either in formal sessions or via informal one-on-one appointments. At one site, after the system was somewhat established, "we were sort of on-site all over the place doing ... point-of-care training, instead of classroom training." Implementation of a CPOE system brings up myriad training-related issues:

- "When we brought the inpatient [portion of the CPOE system] up, we timed it so we would get a new crop of residents so that we would do that 1-hour training with that new crop of residents."

- "We're all expected to be facile in resident training."

- "I can tell you why I became a [support person]. I like to teach, and there's a tremendous amount of teaching that goes along with the position."

In addition to training users, bridgers test the systems:

When any new change comes in, it's put into the test account, and the [support staff] and others who are designated testers hammer on it in the test account.

12.6.4 Skills and Training for Bridgers

At the sites participating in our study, most informants felt strongly that bridgers needed to have clinical backgrounds. The position "does require quite a bit of knowledge of the medical world. You have to know what you're working with." From another clinical support person:

I can't understand how anybody can do this kind of thing without having a medical background. We do have some very highly technical people here who don't have a medical background, and we always have to explain to them, you know, this is urgent, and this will never work this way because that's not the way they think.

Because bridgers often need to "go in and reenter orders," they need to be licensed to do so:

Some of the examples would be the difference between having a clinician have to completely rewrite a set of orders because of some problem and having a clinical [support person] who understands nuances being able to resurrect them or rescue the orders and avoid having to rewrite it.

As one interviewee stated simply, "In this job, you have to have been a user; you have to have a medical background of some sort."

None of the bridgers we interviewed said that they had previous formal training when they were hired into these positions. Instead, they gain skills on the job:

We're making it happen as we go along, we really are . . . I remember when I first started here, I kept asking, "What kind of courses can I take?" There are no courses that you can take for medical information systems, there really aren't . . . sometimes what they [the users] want is not necessarily what they need, so you need to know what it is—to have been there—to know what it is they're talking about.

Another said, "The learning curve was—oh, it was the steepest learning curve I've ever experienced."

I had sort of grown into this local expert role on the computer . . . X really took me under her wing and trained me . . . when I interviewed, I said I don't have the computer skills; I have a very good working knowledge of the system, I know the staff, I know ordering, I know patient care, I know presentation communication—I have all those skills.

Another interviewee said:

The main question that X asked me when I interviewed was how did I diffuse anger, because people get very angry. I don't know that I fully understand why, except that it might have something to do with the loss of control, and people who are used to sort of being captains of the ship being in situations where they don't really know or understand—that's very frightening to them.

In fact, one observer noted:

I was struck by the calm and patient demeanor of the [support staff]. The nurse practitioner was clearly at the end of her fuse, expressing her frustration at every opportunity.

12.6.5 Key Attributes of Bridgers

Patience, assertiveness, and tenacity ranked high on the list of attributes cited as necessary to serve as an effective bridger. One support staff member said:

Once I snag physicians here—and sometimes they're resistant to this—I say, "Look, give me an hour." Then they want to know just a little bit more. And I say, "It's a good investment of your time. Trust me."

In addition, willingness to take the initiative was cited numerous time as a needed attribute for bridgers. We were told that one staff member who lacked initiative did not survive:

X really didn't take a lot of that initial initiative that one had to do to learn how to be a [clinical support person]. He left shortly thereafter.

Unfortunately, burnout is also a factor among these staff members, although many were also long-timers at the organization:

That's a very big service—very diverse and difficult for one person to handle alone. So I think there was a burnout factor there.

12.7 Discussion

This chapter has identified three categories of special people as important and necessary if a CPOE implementation process is to be successful. The sites studied were successful in that CPOE was routinely used by most physicians. The kinds of people described earlier in this chapter were present at each site, and the quotes included are representative of many strong statements made by interviewees and numerous descriptions in field notes.

Lorenzi and Riley[5] provide general insight into why special people are needed; they refer to them as "the cast of characters". These authors reason that people in different stakeholder categories play different roles in the change management process, and they warn that identification of individuals and their roles is important so that a variety of people can be involved from the beginning. Patel and Kaufman[6] offer a line of reasoning based on cognitive science: "There is a need for bridging disciplines to enable clinicians to benefit from rapid technologic advances." As these authors point out, any discipline that is made up of people from different backgrounds—including the field of informatics—needs to find an effective way to communicate, because "many of us are not native speakers of medical informatics". Until the discipline matures enough to have its own language, people who span or bridge disciplines can serve as interpreters.

Each of these categories of special people has been described to some extent in the literature. Top-level support has been recognized as necessary by some, for example, but results of empirical studies have been mixed. Cooper and Zmud found that senior-level support was needed for information technology implementation success.[7] Weir found in a Delphi study that top-level support and commitment are important for CPOE implementation,[8] although in a later survey this factor was not found to be significant.[9] Ash determined that top-level support was not a significant predictor of success for computer-based patient record implementation.[3] The need for champions for any change has been discussed in the literature,[10] with Ash finding that champions are needed for successful infusion of some healthcare information technologies.[3] Weir found that interdisciplinary implementation groups are important in addition to intensive support.[8] Aydin and Forsythe,[11] in their ethnographic study in ambulatory care, found that "tutors available on-site to answer questions in the clinical setting"—the people we are calling bridgers—are key factors in ensuring the successful implementation of the electronic medical record.

The present study provides greater insight into the three categories of special people than has been described previously, however. A number of attributes are shared among the three groups of special people. For example, they are all excellent communicators. The leadership-level individuals connect with one another and with those in the clinical realm. Those in the clinical realm—even the curmudgeons—are connected with the leadership level and with the bridgers. The bridgers themselves are connected with the clinical realm and the CMIO. Bridgers share an interest in interpreting the language and culture of the different groups, thereby inspiring them to act as the interpreters needed by each group.

All levels share a vision, a goal, and a commitment to make CPOE and clinical systems work on behalf of patient care: Even the curmudgeon can be motivated by the promise of an increased ability to provide better patient care. The members of each group follow through with their passion for reaching the goals by utilizing behavior that encourages others: The leadership gives support, the clinical realm provides encouragement to peers, and the bridgers provide "help at the elbow" for users.

Another attribute they share is toughness—an essential quality given that complaints registered by users at the highest levels of the organization and frustrations with the system are often taken out on the individuals who provide support.

The data gathered in our study also raise a number of concerns about these special people. All of these positions are extremely difficult for the people who fill them. All of the individuals wear more than one hat, and their different roles often pull them in different directions. The time commitments required to fulfill these responsibilities effectively are extraordinary. Those who continue to perform clinical duties while taking on responsibilities related to clinical systems are especially vulnerable to overwork. In our study, the prospect of burnout was overtly mentioned in relation to bridgers, but undoubtedly it is a factor in the turnover of staff in other areas as well.

The methodology used for this study was labor intensive, but the complexity of the overarching research question demands the use of multiple perspectives and research methods. The use of observation can verify what one is told in interviews, and group interviews and focus groups offer the benefits of synergy and the building of ideas. The analysis process is time-consuming; nevertheless, to gain a true picture, a mix of researchers with different clinical and research backgrounds must become immersed in the data for a period of time. Studying one site rather than four sites for a shorter period of time with fewer researchers would have produced less data to analyze, but might have produced less transferable results.

12.8 Conclusion and Recommendations: New Bottles, New Wine

The results of this study have implications for informatics education and training. We were told that preparation for the "special people" roles is inadequate at this time. At the CEO level, both clinical knowledge and information systems expertise are often lacking. The CIO may have limited administrative and clinical knowledge. The CMIO seems to need it all. The clinical realm, with the possible exception of the curmudgeons, includes many individuals desiring more information systems expertise. The bridgers are the most extraordinary group: They have clinical expertise, but must gain their technical knowledge on the job. We would recommend that those involved in graduate programs, including programs in health administration and medical informatics, learn more about the roles of these special people so that they can prepare students to fill them.

The results of our study also have budgetary implications. There must be general recognition that these special people are important and that ensuring their presence takes resources, both to hire them and to retain them. Most hospitals have CIOs, and many larger hospitals are hiring CMIOs, but special people in the clinical realm need to be identified as well. Also, a reward system that can motivate the champions, opinion leaders, and even the curmudgeons can increase involvement, thereby improving the chances of success for the CPOE system. It is imperative that bridgers be identified, trained, and rewarded in adequate numbers so that users will have help at the elbow when and where they need it.

The special people have a unique capacity to interpret the hidden cultural assumptions in CPOE and in clinical systems. They must be heralded as the heroes of any successful implementation.

Questions to ask regarding "special people" issues in relation to CPOE include the following:

1. Is there a single, clearly identified CPOE project leader with realistic attitudes about what can and cannot be accomplished, knowledge of how to educate administrators, and skills to foster teamwork?

2. Is there a clinical information system safety committee (could be a subset of CIS oversight committee) that reviews all user reports of system problems and errors resulting from use of the system?

3. Is use of CIS consultants carefully planned with specific objectives before they are employed (if at all)?

4. What percentage of regularly scheduled clinical information system oversight meetings have at least one practicing physician (e.g., a physician who uses the CIS to review and record patient information) and one nursing representative present?

5. Is CPOE one of the top three priorities for the hospital/organization?

6. Is there executive accountability for major clinical information system projects?

7. Does the CDS oversight committee have clinical representation?

8. Is at least one person in the organization with advanced training in clinical informatics (MS or PhD) working with or on the CIS oversight team?

9. Are there employees with extensive clinical experience (e.g., RN, PharmD, MD, lab technician) in the information technology department?

12.9 References

1. Forsythe DE. New bottles, old wine: hidden cultural assumptions in a computerized explanation system for migraine sufferers. *Med Anthropol Quarterly*. 1996;10:551–574.

2. Ash JS, Gorman PN, Lavelle M, et al. Perceptions of physician order entry: results of a cross-site qualitative study. *Methods Inform Med*. 2003;42:313–323.

3. Ash JS. Organizational factors that influence information technology diffusion in academic health sciences centers. *J Am Med Inform Assoc*. 1997;4:102–111.

4. Rogers EM. *The Diffusion of Innovations*. 4th ed. New York: Free Press; 1995.

5. Lorenzi NM, Riley RT. Managing change: an overview. *J Am Med Inform Assoc*. 2000;7:116–124.

6. Patel VL, Kaufman DR. Medical informatics and the science of cognition. *J Am Med Inform Assoc*. 1998;5:493–502.

7. Cooper RB, Zmud RW. Information technology implementation research: a technological diffusion approach. *Manage Sci*. 1990;36:123–139.

8. Weir C, Lincoln M, Roscoe D, Turner C, Moreshead G. Dimensions associated with successful implementation of a hospital-based integrated order-entry system. *Proc AMIA Annu Fall Symp*. 1994;653–657.

9. Weir C, Lincoln M, Roscoe D, Moreshead G. Successful implementation of an integrated physician order-entry application: a systems perspective. *Proc AMIA Annu Fall Symp*. 1995;790–794.

10. Howell JM, Higgins CA. Champions of change: identifying, understanding, and supporting champions of technological innovations. *Organiz Dynamics*. 1990;40–55.

11. Aydin CE, Forsythe DE. Implementing computers in ambulatory care: implications of physician practice patterns for system design. *Proc AMIA Annu Fall Symp*. 1997;677–681.

13 Assessment of the Anticipated Consequences of CPOE Prior to Implementation

Based on: Sittig DF, Ash JS, Guappone KP, Campbell EM, Dykstra RH. Assessing the anticipated consequences of computer-based provider order entry at three community hospitals using an open-ended, semi-structured survey instrument. *Int J Med Inform.* 2008;77(7):440–447.

13.1 Key Points

- Using an open-ended survey may prove useful in helping computer-based provider order-entry (CPOE) leaders to understand user perceptions and predictions about CPOE because it can expose issues about which more communication or discussion is needed.

- Only a few clinicians participating in our study predicted more than a small proportion of unintended or unanticipated events, emotions, and process changes that are likely to result from a CPOE implementation.

- Using the implementation strategies and management techniques outlined in this paper, any chief information officer (CIO) or chief medical information officer (CMIO) should be able to make the necessary midcourse corrections, and be prepared to deal with the unintended consequences of CPOE should they occur.

13.2 Introduction

After extensive reflection on the "unintended" or "unanticipated" findings described in detail in previous chapters of this book, and on the heels of many subsequent conversations with various clinical informatics experts, we began to question whether the "unintended" or "unanticipated" consequences that we had identified were perhaps already known and expected by others in the field and routinely communicated throughout their organizations prior to any computer-based provider order-entry (CPOE) implementations. In an attempt to determine what "average" clinicians were expecting to occur in organizations that were preparing to implement CPOE for the first time, or in organizations that were about to experience a significant system upgrade, we conducted open-ended, semi-structured interviews with clinicians at three community hospitals. This chapter reports on the results of these interviews.

13.3 Background

13.3.1 Rapid Ethnographic Assessment Methods

We developed a short interview guide that we customized (e.g., changing the name of the organization and the names used to refer to their clinical information system) for each site. The survey, with questions based on the categories of unintended consequences we had previously identified, was a form of a rapid behavioral survey of key stakeholders.

13.3.2 The Survey Sites

The following sections describe each of the three hospitals we surveyed. Each hospital was at a slightly different stage in the CPOE implementation pathway. One organization had begun rolling out its CPOE system on one small clinical unit. A second organization was in the early, pre-CPOE phase (e.g., admit–discharge–transfer [ADT], billing, and patient tracking, but no CPOE) of its rollout. The third organization was within 2 months of its planned go-live date. All of the clinicians at each of these sites had received extensive exposure to various forms of clinical information systems over the past several years, in addition to being subjected to an extensive pre-CPOE communication program. All sites were implementing high-quality,[1] full-featured, integrated, commercially available CPOE systems (e.g., CPOE systems created by Epic Systems, Madison, Wisconsin; McKesson, San Francisco, California; and Eclipsys, Atlanta, Georgia).

El Camino Hospital, Mountain View, California

El Camino Hospital is a 399-bed community hospital serving the heart of the "Silicon Valley." The hospital first implemented the Lockheed system (then Technicon Data Systems, Alltel, and now Eclipsys) in the 1970s[2] and has the longest continuously operational CPOE system in the world. At the time of the survey described in this book, the organization was less than 2 months away from a planned hospital-wide upgrade to the new Eclipsys Sunrise system (version 3.5).[3] This upgrade from a 30-year-old, character-based (40 characters × 22 lines per screen), conversational-style interface in which all potential user-selectable options are clearly visible on each screen will move the hospital to use of a state-of-the-art, Windows-based application with a high-resolution (1024 × 768 pixel display) graphical user interface that utilizes many of the currently available data entry and display widgets (e.g., drop-down lists, check boxes, nested hierarchical menus).

Kaiser Permanente, Sunnyside Hospital, Portland, Oregon

Kaiser Permanente's Sunnyside Hospital is a 196-bed community hospital serving Kaiser Permanente Northwest members in the greater Portland metropolitan area. While most of the clinicians at this hospital have extensive experience with its highly successful ambulatory care clinical information system (Epic Systems, Madison, Wisconsin)[4], the only clinical application in use in the hospital is an older, mainframe-based clinical results review application that relies on a character-based menu interface to provide clinicians with access to a patient's clinical laboratory and radiology results. At the time of this survey, Sunnyside Hospital had just "gone live" with phase I (ADT, new inpatient pharmacy, hospital billing, and emergency department tracking systems, all from Epic Systems) of its inpatient clinical information system rollout and was approximately 4 months away from the planned hospital-wide rollout of the CPOE system.

Providence Portland Medical Center, Portland, Oregon

Providence Portland is a 483-bed, community hospital serving the Portland metropolitan area. At the time of the survey, it was approximately 4 months into a CPOE pilot on its rehabilitation unit, with plans to begin a phased rollout of the CPOE system to the rest of the inpatient clinical units over the next 2 years. This organization is implementing McKesson's Horizon Expert Orders CPOE system (McKesson, San Francisco, California) that is based on the system developed at Vanderbilt University over the past 10 years.[5]

13.4 Methods

We created an open-ended, semi-structured interview survey template that we customized for each organization. This survey was designed to be administered orally to clinicians and take approximately 5 minutes to complete, although we did not stop any clinician from discussing the topics in greater depth, if they desired. We administered the survey to clinicians within each organization at common gathering places—a true "convenience" sample.[6] For example, at Kaiser Permanente, we approached clinicians in the cafeteria. At El Camino Hospital, we interviewed clinicians in their computer training facility. At Providence Portland, where many of the clinicians are community-based, a member of the clinical information system staff helped us identify clinicians in and around the cafeteria, as many were in regular "street" clothes rather than the "white coats" and "scrubs" that we encountered at the other institutions. All survey responses were transcribed for data analysis.

All members of the POET research team participated in data analysis. We analyzed each set of surveys from an organization independently and created a summary of the responses. These summaries were fed back to the principal investigator at each site with two purposes in mind: (1) to help him better understand his organization's current state of CPOE anticipation, and (2) to provide member checking[7] to validate our interpretations. We combined all of the site-specific summaries to help us make more general—and ideally transferable—statements about average clinicians' anticipation of the consequences of CPOE.

13.5 Results

13.5.1 The Survey

We interviewed a total of 83 clinicians at the three sites: 31 physicians, 31 nurses, and 21 allied health professionals. There were no major differences in the interviewees within professional categories based on clinical training, years of professional experience, age, or gender. In addition, all of the clinicians at each site were familiar with basic clinical computing features and functions such as patient lookup, clinical results review, email, and Internet access. To preserve the anonymity of the interviewees from each site and because the answers to the questions were fairly consistent across sites (e.g., there were not any significant themes that were observed at only one site), we report the answers to each question in aggregate in this section.

All but two respondents had at least heard about the upcoming CPOE implementation at their organization. Less than one-fourth (25%) of the interviewees had tested or been trained on the systems. In general, interviewees recognized that the new CPOE systems would initially slow them down because there would be a "learning curve" associated with the new system. There were very few openly hostile or negative comments about the impending system implementation.

When asked how the new system might compare to the current system, almost all respondents were able to articulate multiple perceived advantages, including increased legibility, reduced time to find charts, less paper, improved communication, and an overall improvement in patient safety. Most interviewees were also able to describe multiple disadvantages of the new system, including the following concerns:

- More difficult to use (e.g., "more work/new work" from our list of unintended adverse consequences)
- A long learning curve ("never-ending demands")
- More frustration ("emotions")
- Worries about technical issues such as downtime procedures ("overdependence on technology")

When asked about the perceived effect of the new system on other clinicians within the organization, the interviewees' responses varied from cautiously optimistic to quite pessimistic. The comments ranged from anticipated improvement in communication among departments ("communication") to concerns about more work ("more work/new work") to worries about physicians taking their frustrations out on nurses ("emotions/power"). One respondent even mentioned a potential "showdown" between physicians and nurses ("power").

In response to the question about the perceived effect of the new system on patients and patient care, many interviewees recognized that any patient effect would result from the effects or process changes experienced by the clinicians rather than direct effects experienced by the patients. Overall, interviewees thought the new CPOE system would improve patient care following the difficult implementation period. Specifically, they mentioned that care would be more streamlined, faster, and less error prone.

When asked about the perceived effect of the system on the organization as a whole, many interviewees stated that it would improve the perception of the organization in the community and perhaps save money for the organization. Another broadly held opinion was that the impending change was inevitable based on their experiences in other information-intensive service industries (e.g., retail sales, restaurants, airline travel, banking) and that healthcare institutions could no longer defer the adoption of this new technology if they are to remain competitive.

Our final question regarding "what the future of CPOE at each particular organization might look like" was difficult for respondents to separate from our earlier questions. Respondents simply reiterated answers from earlier questions; therefore, we were not able to answer this question.

13.5.2 The Report to CPOE Leadership

Following completion of the analysis of the findings from the surveys, we created and presented an individualized report to each of the three sites. This report served as a form of internal validation of our findings, in addition to informing the CIOs about the state of preparedness of the clinicians within their organizations. The overall state of preparedness of the clinicians at each site was judged based on a consensus of the subjective opinions of the POET research team. This consensus was reached during a debriefing session immediately following the multi-day site visit, during which we administered and then analyzed the survey data, interviewed many clinicians, and carried out many different participant observation sessions.

13.6 Discussion

13.6.1 Survey Results

While our study participants were able to describe both advantages and disadvantages of their impending CPOE rollout, in only a few instances did these clinicians mention any of the "unintended" or "unanticipated" consequences that we have identified in our previous research. Based on our small convenience sample of clinicians at these three community hospital sites, it appears that clinicians recognize that during the transition period, many work processes will take longer and that there will be some new work on their part as they become accustomed to the new system. None of our interviewees anticipated that the new system would continue to slow them down after the initial learning period (as we and others[7,8] have found at multiple CPOE implementation sites).

Many clinicians were worried about how they would take care of patients if and when the system went down. This fits into our "overdependence on technology" unintended consequence category. No one mentioned that the system might cause "new kinds of errors," which we and others[3,4,5,7] have seen and documented at many different CPOE implementation sites.

Further, none of the interviewees mentioned that the new system might have a detrimental impact on communication; in fact, many thought that it would improve communication. While we believe that certain types of communication are improved with the implementation of CPOE (e.g., interdepartmental communication of orders), we have also seen significant negative effects on clinical communication from the following sources:

- Clinicians remotely entering orders that are not completely understood or acted upon in a timely manner
- Changes in clinical decisions following rounds that are not clearly explained or clearly communicated by the new orders
- Failure on the part of nurses to recognize that "new" orders are present for a certain patient

Only one person from the three organizations anticipated any changes in the power structure of the organization owing to the implementation of the CPOE system. In our previous work, however, we have observed many instances in which "CPOE enables shifts in power related to work redistribution and safety initiatives and causes a perceived loss of control and autonomy by clinicians."[9]

In summary, the clinicians we interviewed had a realistic view of the impact that CPOE would have on them and their peers. They understood both the upsides and the downsides of CPOE and even seemed to have a long-term perspective, acknowledging that the system could generate a positive payoff at the end of the long learning curve. They did not indicate that they were aware of many of the unintended consequences of CPOE implementation that we have identified, but they were surprisingly well-informed about CPOE in general.

13.6.2 Study Limitations

As compared to a highly powered, randomly controlled clinical trial, there are too many limitations associated with the present study to mention. Nevertheless, when an entire field is at a very early stage in our understanding of the basic scientific and organizational questions that we must address if we are to improve our implementations of such complex, sociotechnical projects as CPOE, then these limitations can be placed in their proper perspective. For example, the three study sites were selected purely due to chance: We were looking for healthcare organizations that were about to go live with new CPOE implementations. Likewise, our decision to use a convenience sampling strategy within each organization was based solely on our need to identify quickly a relatively small group of clinicians to interview. Both of these decisions may have resulted in a biased sample of clinicians, but without spending an inordinate amount of time and effort to ensure a truly representative sample of clinicians, we could not identify a better approach for conducting this research.

13.6.3 Report Feedback

The open-ended, semi-structured survey succeeded in helping the clinical information technology leadership within each organization better understand the state of preparedness of their clinicians. The CPOE leaders to whom we communicated the survey results were very positive and appreciative of our work. All of them were relieved that their pre-implementation communication strategies were, for the most part, very successful. By "successful," they meant that clinicians throughout the organization had received and understood their messages regarding the impending CPOE implementation.

These organizational leaders had previously sensed that their clinicians had a "realistic appraisal" of the new CPOE product (e.g., they held neither overly positive nor overly negative feelings about the system), but our study results validated their feelings. In addition, they were glad that clinicians recognized that implementation of the system would require a difficult transition period that would improve rather quickly (at least, they hoped it would). Implementation leaders were most concerned that something dramatic might happen leading to a "massive outcry" that might "stop the project right away."

On the one hand, because predicting the future effects of technology on human behavior is a notoriously difficult problem,[10] and because one should never assume that something that has happened in the past cannot happen again,[11,12] the results of this survey were somewhat comforting to the leaders. On the other hand, we found very little evidence that either the clinicians interviewed or the CIOs were preparing for, or even aware of the possibility of, the vast majority of the unintended consequences that we have identified repeatedly during our work at multiple organizations with long-standing CPOE implementations.[13] This finding led us to develop the recommendations presented in the final section of this chapter.

In one organization, clinicians were fearful that not enough computers would be available. In that case, we discussed with the CIO what we felt was a need for far more computing devices in the clinical areas. We were reassured that more devices had been ordered and were scheduled to arrive and would be available before the go-live date. In response to this reply, we suggested that the CIO try to communicate this message to the clinical staff as a way of reducing their anxiety. At another site, the organization decided to delay its CPOE implementation soon after our results were in, although it did not appear that the report of our survey results played a major role in that decision.

13.6.4 Recommendations to Increase Awareness of the Unintended Consequences of CPOE

After further discussions among POET team members, participants from the Menucha retreat,[14,15] and additional discussions with our key contacts at each of the three study sites, we developed the following recommendations. These recommendations are designed to help CPOE implementers increase awareness of, and begin putting into place, the necessary strategies to help their organizations overcome, or at least ameliorate, the unintended consequences that we have previously identified.

Develop and execute a sound organizational change management plan with the help of clinicians.[16] "The level of change involved with CPOE is much greater than hospitals have faced in the past, and success requires that the organization become expert at accomplishing and sustaining change."[17]

Be willing to learn about and address issues of unintended consequences that have occurred elsewhere.[17] In addition to reading as many of the reports about CPOE implementations that have appeared in the literature as possible,[18] leaders of organizations that are interested in implementing CPOE should make a few site visits to organizations similar to theirs that have successfully implemented a similar CPOE system. More often than not, specific lessons can be learned from these projects.[19]

Clinicians must be deeply involved in the CPOE system selection and implementation process. CPOE must not be a project that is perceived as being led by the information technology department. Rather, it must be a clinically driven project chosen to meet one or more of the organization's highest-level strategic goals—for example, to improve patient safety or overall quality of care.[20] If the project is not clinically driven, it will be difficult to obtain enough clinician input to allow the organization to avoid many of the workflow-related unintended consequences.

Communication about the upcoming CPOE implementation must come from individuals who can speak the languages of both technology and clinical workers. Hire good people with knowledge and experience in clinical, information technology, project management, and clinical informatics areas; if possible, hire people who also have experience implementing these types of complex clinical information systems in other organizations.[21] Without input and participation from experts in all of these areas who can talk to the users, many readily identifiable and preventable consequences will arise that have the potential to derail the entire project.

Provide adequate opportunities for all clinicians to be trained in and to experience using the new clinical information system before its deployment. Many organizations require that all clinicians complete a minimum amount of training before they are authorized to log into the live system and enter an order. While this policy may seem rather draconian, usage of new systems is especially problematic for clinicians who have not received formal training in use of CPOE. In addition, each clinical unit should conduct multiple walk-throughs (e.g., dress rehearsals) with all key clinical and administrative personnel involved in all key clinical and administrative processes in which the CPOE system will play a role. For example, physicians should practice writing orders on a newly admitted patient or create the orders for discharge medications; nurses should practice acknowledging the existence of each new order and creating a work list or charting their actions using the medication administration record.[19]

Develop and test downtime and system reactivation procedures. The act of thinking through what will be required to continue to care for patients when the computer system is unavailable will help clinicians to see the enormity of the workflow changes that will occur following CPOE implementation.[22] Assure users that processes will be in place so that system downtime can be well-managed.[23]

13.6.5 Update on the Sites Studied

Over the past several months, Providence Portland Medical Center has experienced a successful CPOE rollout; it has activated additional clinical units without any major setbacks. El Camino Hospital has acknowledged some new medication errors due to problems with its order verification and audit procedures and general unhappiness among the clinical staff with the system.[24] The CPOE rollout at Kaiser Permanente Sunnyside was delayed until late 2008 for financial reasons but was successfully completed.[25]

13.7 Conclusion

The open-ended, semi-structured survey instrument proved useful in helping CPOE leaders to understand user perceptions and predictions about CPOE, and it also exposed issues about which more communication is needed. Our findings did not reveal any overly negative, critical, problematic, or striking sets of circumstances at any of the three organizations, which greatly relieved all of the CPOE site leaders. From the standpoint of our list of unintended adverse consequences, however, we found only a few clinicians who predicted more than a small proportion of unintended or unanticipated events, emotions, and process changes that are likely to result from their CPOE implementation. Using the survey, implementation strategies, and management techniques outlined in this paper, any CIO or CMIO should be able to adequately assess his or her organization's CPOE readiness, be able to make the necessary midcourse corrections, and be prepared to deal with the currently identified unintended consequences of CPOE should they occur.

13.8 References

1. *KLAS CPOE Digest 2007.* Available at: http://www.klasresearch.com/Klas/Site/News/NewsLetters/2007–03/CPOE.aspx. Accessed July 19, 2009.
2. Barrett JP, Barnum RA, Gordon BB, et al. Final report on evaluation of the implementation of a medical information system in a general community hospital. Battelle Laboratories (NTIS PB 248 340), December 19, 1975.
3. Childs B. El Camino turns the page: a pioneering hospital recently exchanged the world's first medical information system for a next-generation solution. *Healthcare Informatics.* August 2006;56–57.
4. Chin H, Brannon M, Dworkin L, et al. Kaiser Permanente–Northwest. In: Overhage JM, Ed. *Proceedings of the 4th Annual Nicholas E. Davies CPR Recognition Award of Excellence Symposium.* CPRI, Bethesda, MD. Chicago: Healthcare Information Management and Systems Society; 1998, pp. 55–100.
5. Miller RA, Waitman LR, Chen S, et al. The anatomy of decision support during inpatient care provider order entry (CPOE): empirical observations from a decade of CPOE experience at Vanderbilt. *J Biomed Inform.* 2005;38(6):469–485.
6. Gliner JA, Morgan GA. *Research Methods in Applied Settings: An Integrated Approach to Design and Analysis.* Lawrence Erlbaum Associates, Mahwah, New Jersey; 2000. p. 155.
7. Patton MQ. *Qualitative Research and Evaluation Methods.* 3rd ed. Thousand Oaks, CA: Sage; 2002.
8. Poissant L, Pereira J, Tamblyn R, Kawasumi Y. The impact of electronic health records on time efficiency of physicians and nurses: a systematic review. *J Am Med Inform Assoc.* 2005;12(5):505–516.
9. Ash JS, Sittig DF, Campbell EM, et al. An unintended consequence of CPOE implementation: shifts in power, control, and autonomy. *Proc AMIA Fall Symp.* 2006; 11–15.

10. Tenner E. *Why Things Bite Back: Technology and the Revenge of Unintended Consequences.* New York: Knopf; 1996.

11. Massaro TA. Introducing physician order entry at a major academic medical center: I. Impact on organizational culture and behavior. *Acad Med.* 1993;68(1):20–25.

12. Morrissey J. Harmonic divergence: Cedars-Sinai joins others in holding off on CPOE. *Modern Healthcare.* 2004;34(8):16.

13. Ash JS, Sittig DF, Poon EG, Guappone K, Campbell E, Dykstra RH. The extent and importance of unintended consequences related to computerized physician order entry. *J Am Med Inform Assoc.* 14(4):415–423.

14. Ash JS, Stavri PZ, Kuperman GJ. Consensus statement on considerations for a successful CPOE implementation. *J Am Med Inform Assoc.* 2003;10(3):229–234.

15. Ash JS, Sittig DF, Seshadri V, Dykstra RH, Carpenter JD, Stavri PZ. Adding insight: a qualitative cross-site study of physician order entry. *Int J Med Inform.* 2005;74(7–8):623–628.

16. Anderson JG, Ramanujam R, Hensel D, Anderson MM, Sirio CA. The need for organizational change in patient safety initiatives. *Int J Med Inform.* 2006;75(12):809–817.

17. Stavri PZ, Ash JS. Does failure breed success: narrative analysis of stories about computerized provider order entry. *Int J Med Inform.* 2003;72(1–3):9–15.

18. For a list of CPOE-related articles, see the Bibliography section on www.cpoe.org.

19. Sittig DF, Ash JS, Zhang J, Osheroff JA, Shabot MM. Lessons from "Unexpected Increased Mortality After Implementation of a Commercially Sold Computerized Physician Order-Entry System." *Pediatrics.* 2006;118(2):797–801.

20. Stablein D, Welebob E, Johnson E, Metzger J, Burgess R, Classen DC. Understanding hospital readiness for computerized physician order entry. *J Comm J Qual Saf.* 2003;29(7):336–344.

21. Ash JS, Stavri PZ, Dykstra R, Fournier L. Implementing computerized physician order entry: the importance of special people. *Int J Med Inform.* 2003;69(2–3):235–250.

22. Kilbridge P. Computer crash: lessons from a system failure. *N Engl J Med.* 2003;348(10):881–882.

23. Campbell EM, Sittig DF, Guappone KP, Dykstra RH, Ash JS. Overdependence on technology: an unintended adverse consequence of computerized provider order entry. *AMIA Annu Symp Proc.* 2007:94–98.

24. Tanenbaum M. Hospital vexed by new computer system. *Mountain View Voice,* July 14, 2006. Available at: http://www.mv-voice.com/story.php?story_id=1770. Accessed July 19, 2009.

25. Measuring Care Quality in Our Hospitals, Kaiser Foundation Hospital, Northwest Region, Kaiser Sunnyside. May 2009. Available at: http://members.kaiserpermanente.org/kpweb/Link.do?html=/kpweb/pdf/nw/nw_quality_Sunnyside.pdf. Accessed July 19, 2009, p. 1–8.

14 Rapid Assessment of Clinical Information System Interventions

Based on: Ash JS, Sittig DF, McMullen CK, Guappone K, Dykstra R, Carpenter J. A rapid assessment process for clinical informatics interventions. *Proc AMIA Fall Symp.* 2008: 26–30.

14.1 Key Points

- Informatics interventions generally take place in rapidly changing settings where many variables are outside the control of the evaluator.

- Assessment must be timely so that feedback can instigate modification of the intervention.

- A generalizable method of inquiry that can help to rapidly identify and assess a situation is desirable for both research and application purposes.

- At the institutions participating in the authors' study, while the sponsors' assistance was crucially important for initially introducing us via electronic mail to potential interviewees, we also needed an on-site "shepherd"—someone who could walk us to units and provide introductions prior to observing—at each site.

- The rapid assessment process is not intended to replace long-term, more traditional ethnographic fieldwork, but it appears to be highly suitable for assessing the rapidly changing context within which clinical information system interventions exist.

14.2 Introduction

Clinical informatics interventions such as implementation of computerized provider order entry (CPOE) with clinical decision support (CDS) are moving evaluation targets: They are continuously changing, as their software and content are updated.[1] Although the ultimate goal is to improve patient care (which is why most studies of CDS have assessed outcomes[2-4]), these studies do not explain why the systems are successful (or not); likewise, they do not provide feedback for iterative system improvements. Formative evaluation methods using naturalistic designs have rarely been used for clinical information system assessment, yet they can best discover how and why systems are successful or not. Kaplan has noted that "these omissions are impoverishing our understanding of CDSS."[5]

191

Qualitative methods are well-suited to investigating the "why" issues, yet traditional ethnographic approaches involve lengthy periods of fieldwork.[6] We often need answers to evaluation questions quickly—while we still have the opportunity to take action and modify the direction toward which we are heading. This ability to respond appropriately in a timely way is especially important in healthcare informatics, given that patient safety can be threatened by unintended consequences. A generalizable method of inquiry that can help to rapidly identify and assess a situation is desirable for both research and application purposes.

Traditional ethnography takes time because researchers must develop cultural competence and knowledge and develop rapport and trust.[6] Rapid methods use several techniques to expedite this process: Data are collected and analyzed by teams; insiders who know the culture are included as team members, and the focus is quite narrow and problem-oriented. Rapid ethnographic assessment using a mix of qualitative and quantitative methods has been used effectively in the public health arena to develop intervention programs for nutrition and primary health care[7] and HIV/AIDS.[8] Also called quick ethnography or the Rapid Assessment Process (RAP) by some,[6,9] this technique comprises a way of gathering, analyzing, and interpreting high-quality ethnographic data expeditiously so that action can be taken as rapidly as possible. The Rapid Assessment, Response, and Evaluation Project (RARE) has been especially well-documented, with manuals available to guide investigators.[10,11] Another tactic for expediting the process is consistent use of structured tools across field sites at the same time that observation and interviews yield high-quality data.

RAP includes many of the methods we have used in past studies but incorporates others as well.[12,13] It relies on a team research approach, including the cooperation of those inside the organization to be studied as well as the external researchers; streamlines the data collection, analysis, and interpretation processes; involves less time in the field; and provides feedback to internal stakeholders. To do so, RAP depends heavily on triangulation of both qualitative and quantitative data. Tools for data collection include site inventory profiles, ethnography guides, interview question guides, and rapid survey instruments.

14.3 Selection of Methodological Approaches

RAP differs from the methods that we employed in our prior research in several ways:

- Use of a preliminary site inventory profile instrument allows researchers to target their questions and observations.

- The semi-structured interviews are less oriented toward taking an oral history and involve two interviewers.

- Data gleaned from short, structured surveys augment the observation and interview data.

- Observations are more focused and include informal interviewing using planned questions.

This type of ethnographic research can be accomplished in a period of several days to several weeks. For most of the work reported on in this book, we spent 3 intensive days in the field at each site, followed by approximately 1 month of analyzing the data.

14.4 Development of the Field Manual

We began by developing a qualitative research field manual that included the following items:

- A site inventory profile or assessment tool
- An interview guide with a list of questions outlining areas to be covered during formal semi-structured interviews
- A schedule for each site visit that outlined work for the 3-day period
- An observation guide including informal questions
- A field survey

Table 14-1 shows a few sample questions included in the site inventory assessment tool.[14] Table 14-2 identifies the areas covered during the formal semi-structured interviews. The observation guide (which is not shown here) included a list of foci and informal questions designed around the site inventory results. For example, a researcher might notice a clinician interacting with a specific CDS module and ask a series of questions: Can you tell me what you think of this feature? Is this the way you usually use it? What would you like to change? Alternatively, if the clinician did not use a feature that the researcher knew was available, the researcher might ask why it was not used and how it could be more useful.

Table 14-1   Example Issues Covered on a Site Inventory Profile Tool Developed to Study CPOE-Related Clinical Decision Support

Hospital characteristics, such as number of staffed inpatient beds

CPOE system information, such as vendor and time since first unit go-live

Hospital locations with CPOE and percentage of units with CPOE

Order-entry system attributes, such as availability of different types of medications and therapeutics, diagnostic tests, and coded clinical data

Clinical decision support types available, such as subsequent or corollary orders, context-sensitive information retrieval, and order sets

CPOE-related applications available, such as an electronic medication administration record (e-MAR) and bar-code medication administration (BCMA)

CDS-related personnel support, including a chief medical information officer, chief nursing informatics officer, and other key managers

CDS-related organizational support, such as multidisciplinary CPOE/CDS oversight committees

Table 14-2   Example Interview Questions from One of the Authors' Formal Interview Guides

Culture: What seems to be the motivation for CDS? What are the cultural barriers and facilitators here? How have attitudes toward CDS shifted over the years?

Control, autonomy, trust: What is the organizational structure (either formal or informal) that relates the quality and IT groups? How do they relate to clinical staff? What are the clinical priorities? Who sets the clinical priorities? How stable is this staff? Who is on CDS committees and why? How do CDS-related committees interact with one another? How do the committees communicate with users? How have they changed over the years? In your estimation, who holds the power here?

Cognition, emotions: What are the barriers and facilitators to use? What is the training for CDS like? How do clinicians keep up-to-date about CDS? How do people feel about CDS?

Content: Where does the organization get its clinical decision-support logic from? How customized is the CDS, and who does it? How often is the clinical content reviewed? What would motivate this hospital to share its content with others? What was implemented when and why?

Human–computer interface: What are the issues surrounding presentation of CDS to clinicians?

Likewise, the field survey was tailored to each new research site depending on the CDS modules available at that institution and on local names given to different features. Questions covered usage, perceptions of CDS, awareness of a CDS committee, involvement of clinicians with development of CDS, communication about the new CDS, and training and support. This short, structured-interview survey instrument was intended to help us to gather information from a wider range of users than those interviewed or observed.

14.5 Preparation for Site Visits

Experience has taught us that careful preparation prior to entering the field is a time saver in the long run. With the help of a local principal investigator/sponsor, we made appointments for interviews and arrangements for on-site observing well before we arrived at the facility. Sponsors also assisted us in completing the site inventory profile and the institutional review board (IRB) paperwork for each site.

14.6 Subject Selection

14.6.1 Sample of Informants

Informants were purposely selected according to their role and relevant knowledge of the CDS system. They included, for example, chief medical information officers; clinician users including physicians, nurses, and pharmacists; quality assurance staff;

information technology staff members; and in-house vendor staff. For selection of clinicians, we deliberately sought out skeptics as well as champions and average users by asking for suggestions from each interviewee using a snowball technique.

14.6.2 Recruitment

The local sponsor invited each selected informant to participate, and then the external principal investigator followed up with detailed information and scheduling. Informants were given small "thank you" items, such as coffee gift cards, upon completion of the interviews.

14.7 Data Collection

Data collection usually took place over 3 days at each site, although we also conducted some follow-up interviews by telephone. Early on day one, we were given a demonstration of the clinical information system, which was especially useful to help us learn the local jargon related to the systems. Interviews were conducted by pairs of researchers and were digitally recorded. Brief field notes were also written during the interviews, so that some notes could be immediately available for preliminary analysis—the complete interview transcription could take several weeks. Four other researchers were on the hospital floors conducting observations and informal interviews, and a doctoral student was stationed in an appropriate common gathering place (e.g., the physicians' lounge) to conduct the field survey. We conducted team debriefings twice a day so that plans could be continuously modified. With 7 researchers, we did close to 15 formal interviews and 40 hours of observation of individuals or units at each site. We also attended meetings of CIS-related committees at most sites. Each site visit ended with a team debriefing that included the local principal investigator/sponsor. We were able to conduct approximately 15 field survey interviews at each site.

14.8 Data Management

Interviews were transcribed by professional qualitative research transcriptionists. Field notes, which were created manually on site, were expanded and put into electronic form by the researchers. Files were entered into N6, formerly QSR NUD*IST (QSR International, Doncaster, Victoria, Australia).

14.9 Data Analysis

To expedite the data analysis, each researcher listened to assigned recordings of interviews, taking notes about identified foci, and reviewed everyone's field notes. Each researcher was then assigned specific topics to summarize. These topics included user perspectives, administrative perspectives, technology issues, and barriers and facilitators. Case reports were written for each site and comments were solicited from those inside the organizations, generating some changes. These case studies formed the basis for a comparative analysis of data.

The interpretive process was both iterative and flexible. Discussions during on-site debriefings, careful formal data analysis, and "member checking"[15] (a qualitative technique to further establish trustworthiness of results by asking insiders for feedback) provided productive and continuous opportunities for interpretation.

14.10 Results

The site inventory profile results proved tremendously helpful in our site visit planning; this instrument needed little modification over the course of our study. In contrast, the observation guide was modified for each site several times. The formal interview guide also evolved as we learned local terminology for systems and units, and as we made discoveries that we wanted to investigate further. In addition, we needed to make major changes in the field survey when we discovered that the questions were inappropriate based on what we learned about the local context and culture. Our sense is that by triangulating data from this variety of sources and by preparing so carefully for visits, we reached saturation at each of the sites within the targeted time period.

14.11 Lessons Learned about Methods

While the sponsors' assistance was crucially important for initially introducing us via electronic mail to potential interviewees, we also needed an on-site "shepherd"—someone who could walk us to units and provide introductions prior to observing at each site. We were fortunate in gaining the assistance of a skilled, locally well-known and well-liked CIS trainer at each hospital. These individuals were highly familiar with the users and the facilities, had access to on-call schedules, were up-to-date on the CIS employed at the facility, and were trusted by the clinicians. We found that half-hour formal interviews were generally sufficient, that attending committee meetings yielded rich data, and that observing with foci in mind allowed the researchers to gain a sense of the context surrounding CIS usage.

We also found that although RAP techniques are efficient and effective, they take their toll on the researchers during fieldwork. Periods of observation were particularly stressful because researchers were under great pressure to be in the right place at the right time to see relevant activities. Also, the logistics of conducting five interviews a day in different hospital and clinic locations were sometimes complex.

14.12 References

1. Ash JS, Stavri PZ, Kuperman GJ. A consensus statement on considerations for a successful CPOE implementation. *J Am Med Inform Assoc.* 2003;10(3):229–234.

2. Hunt DL, Haynes RB, Hanna SE, Smith K. Effects of computer-based clinical decision-support systems on physician performance and patient outcomes: a systematic review. *JAMA*. 1998;280(15):1339–1346.

3. Kawamoto K, Houlihan CA, Balas EA, Lobach DF. Improving clinical practice using clinical decision-support systems: a systematic review of trials to identify features critical to success. *BMJ*. 2005;330(7494):765.

4. Garg AX, Adhikari NK, McDonald H, et al. Effects of computerized clinical decision-support systems on practitioner performance and patient outcomes: a systematic review. *JAMA*. 2005;293(10):1223–1238.

5. Kaplan B. Evaluating informatics applications: clinical decision-support systems literature review. *Int J Med Inform*. 2001;64:15–37.

6. Beebe J. *Rapid Assessment Process: An Introduction*. Walnut Creek, CA: AltaMira Press; 2001.

7. Scrimshaw SCM, Hurtado E. *Rapid Assessment Procedures for Nutrition and Primary Health Care: Anthropological Approaches to Improving Programme Effectiveness*. Los Angeles, CA: UCLA; 1987.

8. Needle RH, Trotter RT, Goosby E, Bates C, von Zinkermagel D. *Crisis Response Teams and Communities Combat HIV/AIDS in Racial and Ethnic Minority Populations: A Guide for Conducting Community-Based Rapid Assessment, Rapid Response, and Evaluation*. Washington, DC: Department of Health and Human Services; 2000.

9. Handwerker WP. *Quick Ethnography*. Walnut Creek, CA: AltaMira Press; 2001.

10. Trotter RT, Needle RH, Goosby E, et al. A methodological model for rapid assessment, response, and evaluation: the RARE program in public health. *Field Methods*. 2001;13:137–159.

11. Trotter RT, Needle R. *RARE Field Team Principal Investigator Guide*. Washington, DC: Department of Health and Human Services; 2000.

12. Ash JS, Smith AC, Stavri PZ. Performing subjectivist studies in the qualitative traditions responsive to users. In: Friedman CP, Wyatt JC, Eds. *Evaluation Methods in Medical Informatics*. 2nd ed. New York: Springer-Verlag; 2006;267–300.

13. Ash JS, Sittig DF, Seshadri V, Dykstra RH, Carpenter JD, Stavri PZ. Adding insight: a qualitative cross-site study of physician order entry. *Int J Med Inform*. 2005;74:623–628.

14. Sittig DF, Thomas SM, Campbell E, et al. Consensus recommendations for basic monitoring and evaluation of inpatient computer-based provider order-entry systems. Proceedings of the Conference on IT and Communications in Health, Victoria, British Columbia, Canada, February 2007.

15. Crabtree BF, Miller WL, Eds. *Doing Qualitative Research*. 2nd ed. Thousand Oaks, CA: Sage; 1999.

15 Basic Microbiologic and Infection Control Information to Reduce the Potential Transmission of Pathogens to Patients via Computer Hardware

Based on: Neely AN, Sittig DF. Basic microbiologic and infection control information to reduce the potential transmission of pathogens to patients via computer hardware. *J Am Med Inform Assoc.* 2002;9(5):500–508.

15.1 Introduction

Over the past 50 years, various forms of computer-based, information management applications have been developed and deployed in the clinical setting.[1] During this time, many system developers have recognized the benefits associated with having computer hardware in the examination room[2] or at the patient's bedside in the hospital.[3,4] More specifically, both Collen[5] and, more recently, the Institute of Medicine[6] have recognized the importance of having clinicians directly involved in data entry activities at the point of care so as to ensure accuracy and timeliness of the data. In addition, over the past several years, the use of portable computing devices by clinicians in the patient's presence has expanded considerably.[7–9] While the need for and benefits of having computers at the patient's bedside for use by clinicians has been well-studied, little attention has been paid to the potential risks of infection to the patient that these devices might pose. A previous history of transfer of microorganisms from other inanimate environmental objects to patients suggests that the presence of computer hardware in the patient setting needs to be examined for this microbial transfer potential.

This chapter reviews the current literature to determine the potential for transmission of pathogens via computer hardware. It also examines basic microbiologic and infection control procedures that might be used to determine and diminish the risk of microbial transfer to patients.

15.2 Review of Pertinent Microbiologic Concepts and Findings

15.2.1 Steps Preceding an Infection: Basic Definitions and Concepts

Humans are surrounded by a number of microorganisms, most of which are completely harmless and some of which are beneficial and even necessary for our existence. At times, however, our interaction with microbes can lead to an infection. An infection is the result of an interaction between a host (the patient) and a microorganism or some of its products (Figure 15-1). In general, at least four factors—some microbial associated and some host associated—determine whether an infection will occur:

- The number of microorganisms present is a key factor: The more microorganisms present, the greater the chance of an infection.

- The particular armamentarium of virulence factors that the microbe has will influence its ability to cause an infection. For example, a bacterium that produces a particularly potent toxin is more apt to cause an infection than one that does not.

- The most critical factor that the host brings to the interaction is immunologic status. A person who is immunosuppressed or immunocompromised as a consequence of any number of circumstances (Table 15-1) will be more susceptible to an infection.

- For an infection to occur, the microorganism or its products must come in contact with the host.

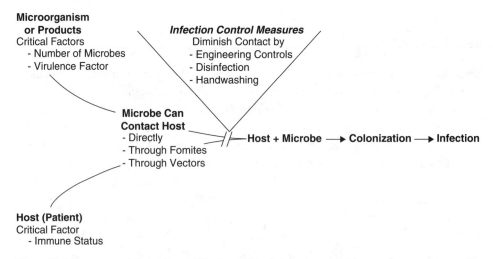

Figure 15-1 Steps potentially leading to infection and basic infection control interventions to decrease the risk of infection.

Table 15-1   Patients at Increased Risk for Infections

Patients of extreme age—either very young or very old
Victims of severe trauma, such as burns or crush injuries
Malnourished patients
Patients with immunocompromising diseases (e.g., diabetes, AIDS)
Patients immunosuppressed for transplantation or chemotherapy

Source: Neely AN, Sittig DF. Basic microbiologic and infection control information to reduce the potential transmission of pathogens to patients via computer hardware. *J Am Med Inform Assoc.* 2002;9(5):500–508.

The last factor, contact, is the focus of this chapter. Contact can happen in a number of different ways. For example, a microbe might directly contact the host, or it might contact the host via an indirect route involving inanimate objects, called fomites, and/or living organisms, called vectors. Once the fomite, such as a piece of computer hardware, or the vector, such as a healthcare worker, becomes contaminated with the microbe, this source then serves as a reservoir for transmitting the microorganism to the host by some form of contact.

Once the microbe reaches the host, a number of different associations are possible. The presence of a microbe in or on a host with growth and multiplication of that microorganism, but without tissue damage, is termed colonization.[10] Once tissue damage begins, the colonization becomes an infection. Not all colonizations become infections, but all infections are generally preceded by colonizations.[11]

15.2.2 Nosocomial Infections: What Are They and Why Be Concerned?

Nosocomial infections are infections that develop within a healthcare institution or are produced by organisms acquired during a stay at such a facility.[12] These infections are not present or incubating at admission but rather are acquired by the patient during some interaction at the hospital or medical unit.

Nosocomial infections cause a significant amount of morbidity and mortality, and they lead to notably increased medical costs. In the United States, an estimated 2 million patients acquire nosocomial infections each year.[13] These acquired infections directly cause approximately 19,000 deaths annually and contribute indirectly to an additional 80,000 deaths each year.[14] Patients who survive nosocomial infections require longer hospital stays with increased medical support. In 1992, the estimated annual cost to treat nosocomial infections was $4.5 billion.[15] To put these figures in a more comprehensible context, a study published from the University of Michigan in 1999 compared the cost of treating patients who had acquired a nosocomial bloodstream infection with the cost of treating patients who had not developed such an infection. Of the patients who survived, the average cost for patients who had picked up a bloodstream infection while in the hospital compared to those who had not was $34,508 more per patient.[16] Hence, these hospital-acquired infections are a concern because of both the human and economic tolls that they exact.

15.2.3 Computers as Microbial Reservoirs: Have Computers Been Linked to Nosocomial Infections?

It has long been recognized that inanimate objects in the patient's environment can harbor microorganisms. These objects might be medical tools, such as stethoscopes,[17] ear thermometers,[18] or bronchoscopes,[19] or common nonmedical objects, such as ballpoint pens,[20] bedrails and bedside tables,[18,21] or plumbing components that introduce microbes into the patient's bath water.[22–24] Only recently have investigators begun to examine the potential for microbial contamination on computer hardware and to ask whether these microorganisms might play a role in hospital-acquired infections.

A search of the literature, using the MEDLINE, BIOSIS, and CINAHL databases, indicated that only a few studies investigating links between computers and patient colonizations and/or nosocomial infections presently exist in peer-reviewed scientific journals (Table 15-2). In 1995, in a letter to the editor, Masterton et al[25] linked refractory methicillin-resistant *Staphylococcus aureus* (MRSA) carrier status in a nurse to contamination of home environmental objects, including a computer desk and a joystick. Despite antibiotic treatment of the nurse, the bacterial carriage of this microbe was not eliminated until the home environment, including computer-related hardware, was decontaminated.

In 1998, a brief report by Isaacs et al[26] described how 27 keyboards in a burn unit were swabbed one time to determine if the keyboards could be contributing to an increase in antibiotic-resistant bacteria in patients residing on the unit. Resistant isolates were not found, leading the authors to conclude that the computer keyboards were not a significant source of the spread of the resistant bacteria in their unit. It is interesting that while the two types of antibiotic-resistant bacteria that they sought were not found, other bacteria—namely, *S. aureus* and *Pseudomonas*, both of which are capable of causing serious infections in burn patients—were found on the computer terminals.

In 1999, Neely et al[27] reported a more extensive study in a burn unit that had experienced an increase in acquired *Acinetobacter baumannii* colonizations. An epidemiologic investigation showed this microorganism to be present more often on computer keyboard covers than on any other objects in the patients' rooms. The increase in patient colonization coincided temporally with the introduction of bedside computers into the patients' rooms. Once control measures were introduced to decrease the presence of microorganisms on the keyboards, the colonization rate for *A. baumannii* on the burn patients returned to its original low level. Such findings strongly suggest a link between contaminated computer keyboards and colonization in this group of patients.

In 2000, in an excellent study in an adult intensive care unit (ICU), Bures et al[28] cultured a number of microorganisms, including MRSA, *Enterococcus*, and *Enterobacter*, from computer keyboards. Cultures from patients in the ICU showed similar microorganisms. In recognition of the fact that MRSA can potentially be a particularly dangerous microbe, the MRSA strain found on the keyboards was compared with the MRSA strain from the infected patients, using pulse-field gel electrophoresis (a particularly sensitive molecular genetics technique for distinguishing among isolates of the same genus and species). This technique showed that the strain causing clinical infection in two of the ICU patients was identical to the strain isolated from the keyboards, thereby establishing a direct connection between the infected patients and the computers.

In 2001, Devine et al[29] cultured for MRSA on ward computer terminals in two different hospitals. In hospital A, 12 terminals were cultured and 5 (42%) tested positive for MRSA. In hospital B, 13 terminals were swabbed and 1 (8%) tested positive for the bacteria. Not surprisingly, hospital A had a significantly higher rate of MRSA transmission for its patients compared to hospital B. These data are consistent with computer keyboards playing a role in the transmission of the bacteria.

Table 15-2 Studies Investigating Computer Contamination and Patient Colonizations or Infections

Year	Author	Study Scope	Primary Findings	Infection Control Measures	
				Before Study	Added After Study
1995	Masterton et al.	Case study, one home computer	MRSA on home computer contributed to MRSA carriage of nurse	None reported	Decontaminate home computer
1998	Isaacs et al.	27 hospital computers tested one time	Antibiotic-resistant microbes sought but not found, but S. aureus and Pseudomonas isolated	Keyboard covers, one time/day disinfection	None reported
1999	Neely et al.	Epidemiologic study of A. baumannii colonization	A. baumannii colonization in patients linked to bedside computer keyboards	Keyboard covers, random cleaning	Daily keyboard disinfection; change in handwashing and gloving policy
2000	Bures et al.	Pulse field gel electrophoresis study of ICU infections	MRSA infections in patients directly linked to computers in ward	None reported	Keyboard covers, disinfected daily; handwashing enforced
2001	Devine et al.	25 terminals cultured one time in 2 hospitals	42% of computers positive for MRSA in hospital A, 8% positive in hospital B; hospital A had higher MRSA transfer rate	Handwashing in both hospitals, but monitored in hospital B	Enforce staff handwashing before and after patient contact
2001	Ivey et al.	Abstract of CPU fan contamination and fungi in patients' rooms	No correlation between isolates on CPU fans and fungi in patients' rooms	None reported	None reported

Source: Adapted from, Neely AN, Sittig DF. Basic microbiologic and infection control information to reduce the potential transmission of pathogens to patients via computer hardware. *J Am Med Inform Assoc.* 2002;9(5):500–508.

To our knowledge, only one other study in the scientific literature has addressed the issue of computers and microbial contamination in relation to patients. In this 2001 study, which has appeared only as a published abstract, Ivey et al[30] asked whether the fan in the computer processing unit (CPU) might be responsible for the dissemination of fungal flora in their ICU. Data showed little commonality between fungal cultures from the CPU dust and cultures from other room areas. The investigators concluded that computer fans in the ICU did not have a significant impact on the fungal infections in their unit.

The introduction of computers into critical care environments is a relatively new phenomenon. Consequently, the full impact of the physical presence of these devices in patient care areas has not been elucidated. However, results from the first early studies clearly demonstrate that the keyboards of computers, just like other bedside inanimate objects,[18,19,22–24] can be reservoirs for microorganisms associated with colonized or infected patients. Whether other computer hardware, such as computer mice, rollerballs, touchscreens, joysticks, or even portable, hand-held devices, might be a factor in the dissemination of microbes remains to be determined.

15.2.4 Microbial Survival and Transfer: Which Factors Influence the Link Between Computers and Patients?

Two factors that play a role in the link between any fomite, such as a piece of computer hardware, and the patient are (1) the ability of a particular microbe to survive on a particular surface and (2) the fact that various vectors, such as healthcare workers, can transfer microorganisms from one surface to another.

Microorganisms survive for different periods of time on different surfaces. Their survival rate varies depending on the particular microbe, the particular surface, and the concentration of the microorganisms on the surface. In general, the greater the concentration of the microbes, the longer they will survive.[31–35] Survival can range from minutes to months. Obviously, if a microbe survives for only a few minutes on an inanimate object, such as the computer terminal, then the possibility of that microbe being acquired by a patient is quite small. Conversely, if particular bacteria or fungi can survive for weeks to months on a certain surface, then the odds of that organism being picked up by a patient or healthcare worker are considerably increased.

Most of the accessible components of computers are made of plastic. In a series of studies in which the survival of a variety of bacteria and fungi on a number of different fabrics and plastics (including the plastic skins used to protect computer keyboards) was quantified, microbial survival was often in the range of days to weeks on both types of surfaces.[33–35] However, when there was a difference in survival between the fabrics and the plastics, the microbes tended to live longer on plastics. Hence, the long survival times of certain microorganisms, particularly on plastics such as those associated with computers, contributes to the possibility of computers acting as reservoirs for these microbes.

How do the microbes get from the surface to the patient? Controlled studies have shown that microorganisms can be readily transferred from inanimate objects to hands, and vice versa. Rangel-Frausto et al[36] found that in 90% of their tests, the yeast *Candida albicans* was transferred from a plastic surface to a person's hands; in 90% of their trials, the yeast was transferred from the hands to a plastic surface. In a more recent study, Noskin et al[37] showed that the bacterium *Enterococcus faecium* likewise was directly transferred from a vinyl surface to a person's hands. Studies have also demonstrated that microbes can be transferred from person to person. In the Rangel-Frausto et al study,[36] the yeast was transferred from hand to hand 69% of the time, and various outbreaks of both bacterial and fungal infections in patients have been traced to a specific individual healthcare worker.[25,38,39] Hence, these studies indicate that it is quite possible for a long-lived microbe on a computer keyboard to be transferred first to a staff member's hands and then to a patient, where it could potentially cause an infection.

15.3 Potential Solutions with Emphasis on Basic Infection Control Principles

All of the solutions presented in this section are quite simple, apply to most inanimate objects, and are based on one principle: Before a microbe or its product can potentially cause an infection in a patient, it must come in contact with that patient. All of the solutions discussed here have the single purpose of decreasing or eliminating the number of computer-associated microorganisms that come in contact with a patient (see Figure 15-1).

15.3.1 Engineering or Process Controls Versus Behavioral Controls

In general, it is preferable to engineer the physical environment or configure a process so that it is difficult for an error, such as contamination, to occur rather than to depend on consistent, meticulous behavior alone to prevent errors (contamination). For example, in many patient rooms, space is at a premium. Given this fact, the computer terminal might be located close enough to the sink so that it could be splattered, and thereby contaminated with microorganisms, during the course of cleaning objects or hands.

To deal with this potential for contamination, one control would be to advise staff to be careful not to splash the keyboard when using the sink; unfortunately, because personnel have multiple duties, it is unlikely that this care would always occur. A better control would be either to relocate the computer or to simply place a water-impermeable barrier, such as a plastic panel, between the sink and the keyboard. With the barrier in place, the behavior of the people using the sink becomes a mute point as far as splashing the keyboard is concerned. Other examples of engineering and process controls might include the use of a computer keyboard cover and of an infrared mouse to allow the process of computer cleaning/disinfection to be easier and more effective than relying on a person to meticulously clean the keyboard or the mechanical mouse without harming the hardware.

Such engineering or process controls may take a little forethought and may involve a bit of expense. Even so, if they save staff time, decrease the need for continuous staff behavior surveillance and education, or prevent nosocomial infections, they are often worth the up-front time and expense.

15.3.2 Cleaning and Disinfecting

Cleaning is the removal of all foreign material, such as dirt and organic material, from an object.[40] Sterilization is the complete elimination or destruction of all forms of microbial life, whereas disinfection is a process that eliminates many or all pathogenic microorganisms, with the exception of bacterial spores, from inanimate objects.[40] Because there are relatively few situations in which computer hardware would need to be sterilized, this discussion basically addresses disinfection, although many of the comments made here are also applicable to sterilization.

Because dirt can harbor microbes from the normal disinfecting process, successful disinfection should be preceded by cleaning. However, certain disinfectant cleaners may accomplish both tasks in one process.

There is no perfect disinfecting agent; each chemical has its own advantages and disadvantages, depending on the situation in which it is used. Therefore, in any medical facility, the infection control personnel should be consulted about appropriate cleaning/disinfecting agents and procedures. Factors to be considered include the level of disinfection necessary for that particular computer, the potential types of organic and microbial contamination that might be present, and the cleaning/disinfecting agents available. When choosing these agents, besides efficacy in disinfection, issues such as patient and personnel safety (e.g., flammability, toxicities), ease of use (e.g., availability, need for pre-mixing), aesthetics (e.g., odors, color changes), and costs need be considered. Guidelines for selection and use of disinfectants are available in what is now a classic article in the infection control literature.[41]

In addition, healthcare workers need to assess the compatibility of the disinfecting chemical with the computer hardware to be cleaned and disinfected. Many chemical disinfectants require that the surface to be disinfected be exposed to the liquid disinfectant for 10 minutes. Such exposure could create an electrical or corrosive problem to certain pieces of computer hardware. With some equipment, such as the computer keyboard, the problem of chemical damage to the keyboard components can be alleviated by the use of a thin plastic keyboard cover (a "skin"), which can be liberally soaked with disinfectant without fear of compromising the computer.

15.3.3 Handwashing and Gloving

Microorganisms on the skin are generally classified into two categories: resident flora and transient flora. Resident flora are microbes that normally colonize or live on the skin of most individuals; they generally do not cause infections unless they are introduced into normally sterile body sites or the host is immunocompromised. In contrast, transient flora are microbes that are present on the skin for only a short time; they tend to be more pathogenic than the resident flora and are responsible for most nosocomial infections.[42] These transient (contaminant) flora may be picked up by the hands of a healthcare worker—for example, when the worker touches a patient or any contaminated object, such as a computer component. Handwashing is a process that removes soil and transient microorganisms from the hands. Hence the simple process of handwashing has long been a mainstay of any control measure for reducing nosocomial infections.

Two basic types of soaps are available for handwashing: soaps that contain an antimicrobial element and soaps that do not include this kind of ingredient. Because of concern about the emergence of resistance to antiseptics, antimicrobial soaps are generally not recommended for regular handwashing,[43] although there may be some specialized areas in a medical facility in which washing with antimicrobial soaps is preferred. Resident infection control personnel can advise about agents for hand hygiene as well as the specific handwashing procedure to use. Important factors include mechanical rubbing of the soap over all surfaces of the hands and an adequate period of rubbing to release the transient organisms.

In addition to soaps for handwashing, "waterless" agents are available for hand hygiene purposes. These alcohol rubs are presently being considered as a replacement for soap and water in the 2002 Guideline for Hand Hygiene developed by the CDC's Healthcare Infection Control Practices Advisory Committee.[44] It is important to realize that these agents are disinfectants—not cleaners. Therefore, any visible soil must first be removed before the alcohol will be completely effective. Also, it is recommended that after five or six consecutive uses, the hands be washed with soap and water to remove any build-up of agent.

A word of caution about gloves: Use of gloves is not a substitute for handwashing. Generally, hands should be washed before gloves are donned; gloves should be picked up by the cuff to prevent contamination of the surface, which may touch a patient or clean object, and hands should be washed after gloves are removed.[45,46] Gloves provide an extra amount of protection; thus they may be used as an adjunct to handwashing but not as a replacement for handwashing. There are certainly circumstances in which gloves can be used to decrease the transfer of microbes,[47] but it is important to note that gloves alone, without an appropriate protocol for their use, could potentially increase transfer of transient flora by giving the wearer a false sense of security. For example, washing one's hands and putting on gloves will prevent the wearer's resident flora from touching the patient or computer and the patient's or computer's microbes from reaching the hands of the wearer. Nevertheless, gloves do not prevent the wearer from transferring microbes from the computer to the patient, or vice versa, because the gloves can carry organisms from place to place or from person to person just as readily as the ungloved hands.

15.4 Practical Applications of Infection Control Principles in Medical Settings

Because each medical facility is somewhat different, the best infection control protocols for each piece of computer hardware or for any other inanimate object in the patient's environment will vary with each situation. Hence, it would be impossible to provide specific protocols for all circumstances. Nevertheless, by applying the previously discussed basic infection control principles, one can address most computer-infection control situations, regardless of which computer component is involved or whether the computers are permanent room hardware or personal hand-held devices. Furthermore, when introducing any piece of computer hardware into any medical situation, the following guidelines might be helpful.

15.4.1 Consult with the Facility's Infection Control Personnel

The infection control personnel may be part of the performance improvement, risk management teams, or have some other designation. Whatever the name used, all medical facilities should have some individual(s) dedicated to general infection control issues. There are several advantages to working with the local infection control staff when considering the application of computer technology.

First, if a new piece of hardware will enter into the patient's area, staff will appreciate knowing this fact because it constitutes a change in the patient's environment. Should any changes in colonization rate or infection rate occur in the patients, the new elements of the environment could then be immediately evaluated to see if they are a contributing factor.

Second, the local infection control staff will be familiar with the routine cleaning and disinfecting agents as well as the routine cleaning schedule. From a practical point of view, if the hospital's cleaning and disinfecting routines are appropriate for the new equipment, these accepted routines should be used rather than a totally different protocol being introduced for one piece of equipment.

Third, if special infection control-related protocols need to be established relative to the hardware, the infection control staff can advise the computer or technical services department how to monitor for compliance with the protocols.

15.4.2 Determine the Risk Level of the Patients Served at Each Computer Hardware Location

Recognizing that microorganisms are ubiquitous and that most microbes are harmless to most people, it would be a waste of both time and money to impose more computer hardware infection control procedures than are needed to protect the patient population. At the same time, being cognizant of the higher morbidity, mortality, and costs associated with nosocomial infections, it is imperative that adequate infection control procedures are in place to protect high-risk patients from these potentially deadly problems (see Table 15-1). Hence, one needs to balance the infection control measures with the level of risk of the patients being served.

For example, in a clinic providing well-child check-ups and immunizations, minimal infection control protocols might be recommended. A keyboard cover and perhaps a monitor cover, if the unit is in striking range of an examination table that might accommodate a male child without a diaper, would be a reasonable investment. The cleaning agents and cleaning frequency for these pieces of equipment could reasonably be the same as those routinely used to clean something like the telephone in the same exam room.

In contrast, in an intensive care unit or a facility serving immunocompromised patients, infection control measures would be much more rigorous. Keyboard covers should be disinfected more frequently than in the well-child clinic, and specific handwashing and perhaps gloving protocols, as recommended in past computer contamination studies,[47] would be appropriate. Exactly how these issues are handled will depend in part on how the computer hardware is being used—the topic discussed next.

15.4.3 Determine How the Computer Hardware is Being Used

The actual usage of the computer component also affects appropriate control measures. Do personnel go back and forth between the computer and the patient? Do staff enter a patient room simply to use the computer and then leave and go to another patient's room? Does the piece of computer equipment move from room to room? In the case of the latter two questions, it is important to remember that anything in the patient's environment (e.g., in the patient's room) will probably be contaminated with microorganisms from that patient. In an intensive care unit situation, it is quite likely that the patient will be colonized by microorganisms that can cause nosocomial infections in other ICU patients. Therefore, anyone (such as a staff member) or anything (such as a portable computer) that contacts anything in the patient's room should be considered to be contaminated and needs to be disinfected before leaving the room.

Consider the following example: A staff person enters an ICU room, washes the hands, and dons gloves, according to hospital protocol; then the person enters data into the hand-held computer and sets the device down in the patient's room, retrieves the hand-held computer, removes the gloves, and washes the hands before leaving the room. Because the hand-held computer contacted a surface in the ICU patient's room, it should be considered to be contaminated with the ICU patient's flora. It also needs be recognized that the staff member, even though he or she followed all personnel protocols for handwashing and gloving upon entering and leaving the ICU, could still transfer microorganisms to the next ICU patient through the hand-held computer device.

Clearly, if a piece of portable computer equipment comes in contact with any of the environment in a patient's room, then that piece of equipment needs to be decontaminated before being brought into another ICU patient's room. If the decontamination process for the piece of portable equipment is complicated, then consideration should be given to restricting the use of these portable devices in rooms of immunocompromised ICU patients.

15.5 Summary

Since the introduction of computers into the healthcare setting 50 years ago, there has been a growing recognition of the value of this technology in providing quality medical care. As the variety of computer devices has increased from PCs to various portable units, and as the availability of software packages has grown from medical records programs to diagnostic aides, there has been an increased presence of computer hardware in all patient care venues—from admissions and clinic areas to ICUs. Only recently have studies begun to investigate whether these computer devices can serve as fomites for the harboring and transfer of microorganisms involved in nosocomial colonizations or infections in hospitalized patients. Considering the long periods of time for which some microorganisms can survive on plastic surfaces and the fact that microbes can be readily transferred from inanimate surfaces to hands, and vice versa, it is not surprising that frequently touched computer keyboards have been implicated in nosocomial colonizations and infections in a variety of patient populations (see Table 15-2).

Infection control measures can be quite simple. For example, they may include engineering modifications, such as the use of keyboard covers, cleaning and disinfection of appropriate computer hardware surfaces, and handwashing with or without gloving of pertinent personnel (see Figure 15-1). When these control practices are used and to what extent they are used will depend on the risk posed to the patient population being served (see Table 15-1) and the practical feasibility (time and cost) of the measures being considered.

As might be expected, different medical facilities have instituted various levels of infection control relative to computer equipment (see Table 15-2). In general, the medical facility's resident infection control staff can advise IT personnel about that facility's routine control practices for medical devices. In conjunction with this staff, specifics for the microbiologically safe use of individual computer hardware in a variety of medical settings can be determined. Observance of these simple control procedures can potentially decrease morbidity and mortality for patients and reduce medical care costs for hospitals and caregiving organizations.

Questions to ask regarding infection control issues in relation to use of computer hardware in the healthcare setting include the following:

1. To reduce the chance of transmission of infectious agents in a routine outpatient clinic, have all computer keyboards and their monitors been covered?

2. Are keyboard covers cleaned as routinely as the telephone in the same exam room and with similar cleaning agents?

3. In an intensive care unit or a facility serving immunocompromised patients, are infection control measures much more rigorous? Keyboard covers should be disinfected more frequently than in the well-child clinic, and specific handwashing and perhaps gloving protocols, as recommended in past computer contamination studies, should be employed.

15.6 References

1. Collen MF. *A history of medical informatics in the United States, 1950–1990.* Washington, DC: American Medical Informatics Association; 1995.

2. Mitchell E, Sullivan F. A descriptive feast but an evaluative famine: systematic review of published articles on primary care computing during 1980–97. *Br Med J.* 2001;322:279–282.

3. Bradshaw KE, Gardner RM, Clemmer TP, et al. Physician decision making: evaluation of data used in a computerized ICU. *Int J Clin Monit Comput.* 1984;1(2):81–91.

4. Wilson D, Nelson NC, Rosebrock BJ, et al. Using an integrated point of care system: a nursing perspective. *Top Health Inf Manage.* 1994;14(4):24–29.

5. Collen MF. General requirements for a medical information system (MIS). *Comput Biomed Res.* 1970;3(5):393–406.

6. Dick RS, Steen EB, Eds. *The Computer-Based Patient Record: An Essential Technology for Health Care*. Washington, DC: Committee on Improving the Patient Record, Institute of Medicine, National Academy Press; 1991.

7. Halpern NA, Burnett G, Morgan S, et al. Remote communication from a mobile terminal: an adjunct for a computerized intensive care unit order management system. *Crit Care Med.* 1995;23(12):2054–2057.

8. Tshopp M, Geissbuhler A. Use of hand-held computers as bedside information providers. *Medinfo.* 2001;10(Pt 1):764–767.

9. Criswell DF, Parchman ML. Hand-held computer use in U.S. family practice residency programs. *J Am Med Inform Assoc.* 2002;9(1):80–86.

10. Jackson MM, Tweeten SM. General principles of epidemiology. In: Pfeiffer JA, Ed. *APIC Text of Infection Control and Epidemiology, vol. 1*. Washington, DC: Association for Professionals in Infection Control and Epidemiology; 2000:1–17.

11. McCormick R. Intensive care unit. In: Pfeiffer JA, Ed. *APIC Text of Infection Control and Epidemiology, vol. 1*. Washington, DC: Association for Professionals in Infection Control and Epidemiology; 2000:45–52.

12. McGowan JE Jr, Metchoch BG. Infection control epidemiology and clinical microbiology. In: Murray PR, Baron EJ, Pfaller MA, Tenover FC, Yolken RH, Eds. *Manual of Clinical Microbiology*. 7th ed. Washington, DC: ASM Press; 1999:107–108.

13. Centers for Disease Control and Prevention. Public health focus: surveillance, prevention, and control of nosocomial infections. *MMWR.* 1992;41:783–787.

14. Maki DG. Nosocomial infection in the intensive care unit. In: Parrillo JE, Bone RC, Eds. *Critical Care Medicine: Principles of Diagnosis and Management*. St. Louis: Mosby; 1995:893–954.

15. Martone WJ, Jarvis WR, Culver DH, Haley RW. Incidence and nature of endemic and epidemic nosocomial infections. In: Bennett JV, Brachman PS, Eds. *Hospital Infections*. Boston: Little, Brown; 1992:577–597.

16. DiGiovine B, Chenoweth C, Watts C, Higgins M. The attributable mortality and costs of primary nosocomial bloodstream infections in the intensive care unit. *Am J Respir Crit Care Med.* 1999;160:976–981.

17. Boyce JM, Opal SM, Chow JW, et al. Outbreak of multidrug-resistant *Enterococcus faecium* with transferable *vanB* class vancomycin resistance. *J Clin Microbiol.* 1994;32:1148–1153.

18. Porwancher R, Sheth A, Remphrey S, et al. Epidemiological study of hospital-acquired infection with vancomycin-resistant *Enterococcus faecium*: possible transmission by an electronic ear-probe thermometer. *Infect Control Hosp Epidemiol.* 1997;18:771–774.

19. Sorin M, Segal-Maurer S, Mariano N, et al. Nosocomial transmission of imipenem-resistant *Pseudomonas aeruginosa* following bronchoscopy associated with improper connection to the Steris System 1 processor. *Infect Control Hosp Epidemiol.* 2001;22:409–413.

20. Datz C, Jungwirth A, Dusch H, et al. What's on doctors' ball point pens? *Lancet.* 1997;350:1824.

21. Slaughter S, Hayden MK, Nathan N, et al. A comparison of the effect of universal use of gloves and gowns with that of glove use alone on acquisition of vancomycin-resistant enterococci in a medical intensive care unit. *Ann Intern Med.* 1996;125:448–456.

22. Weber DJ, Rutala WA, Blanchet CN, et al. Faucet aerators: a source of patient colonization with *Stenotrophomonas maltophilia. Am J Infect Control.* 1999;27:59–63.

23. Berrouane YF, McNutt L, Buschelman BJ, et al. Outbreak of severe *Pseudomonas aeruginosa* infections caused by a contaminated drain in a whirlpool bathtub. *Clin Infect Dis.* 2000;31:1331–1337.

24. Vochem M, Vogt M, Doring G. Sepsis in a newborn due to *Pseudomonas aeruginosa* from a contaminated tub bath. *N Engl J Med.* 2001;345:378–379.

25. Masterton RG, Coia JE, Notman AW, et al. Refractory methicillin-resistant *Staphylococcus aureus* carriage associated with contamination of the home environment. *J Hosp Infect.* 1995;29:318–319.

26. Isaacs D, Daley A, Dalton D, et al. Swabbing computers in search of nosocomial bacteria. *Ped Infect Dis J.* 1998;17:533.

27. Neely AN, Maley MP, Warden GD. Computer keyboards as reservoirs for *Acinetobacter baumannii* in a burn hospital. *Clin Infect Dis.* 1999;29:1358–1360.

28. Bures S, Fishbain JT, Uyehara CFT, et al. Computer keyboards and faucet handles as reservoirs of nosocomial pathogens in the intensive care unit. *Am J Infect Control.* 2000;28:465–470.

29. Devine J, Cooke RPD, Wright EP. Is methicillin-resistant *Staphylococcus aureus* (MRSA) contamination of ward-based computer terminals a surrogate marker for nosocomial MRSA transmission and handwashing compliance? *J Hosp Infect.* 2001;48:72–75.

30. Ivey JC, Oomen B, Forstall G. Fungal contamination related to personal computer devices installed in a hospital intensive care unit. *Am Society for Microbiol Abstrs.* 2001;L-1:469.

31. Dickgiesser N, Ludwig C. Examinations on the behaviour of gram positive and gram negative bacteria on aluminum foil. *Zentbl Bakteriol Hyg Abt I Orig B.* 1979;168:493–506.

32. Watson SP, Clements MO, Foster SJ. Characterization of the starvation-survival response of *Staphylococcus aureus. J Bacteriol.* 1998;180:1750–1758.

33. Neely AN, Maley MP. Survival of enterococci and staphylococci on hospital fabrics and plastic. *J Clin Microbiol.* 2000;38:724–726.

34. Neely AN. A survey of gram-negative bacteria survival on hospital fabrics and plastics. *J Burn Care Rehabil.* 2000;21:523–527.

35. Neely AN, Orloff MM. Survival of some medically important fungi on hospital fabrics and plastics. *J Clin Microbiol.* 2001;39:3360–3361.

36. Rangel-Frausto MS, Houston AK, Bale MJ, et al. An experimental model for study of *Candida* survival and transmission in human volunteers. *Eur J Clin Microbiol Infect Dis.* 1994;13:590–595.

37. Noskin GA, Bednarz P, Suriano T, et al. Persistent contamination of fabric-covered furniture by vancomycin-resistant enterococci: implications for upholstery selection in hospitals. *Am J Infect Control.* 2000;28:311–313.

38. Trick WE, Kioski CM, Howard KM, et al. Outbreak of *Pseudomonas aeruginosa* ventriculitis among patients in a neurosurgical intensive care unit. *Infect Control Hosp Epidemiol.* 2000;21:204–208.

39. McNeil SA, Nordstrom-Lerner L, Malani PN, et al. Outbreak of sternal surgical site infections due to *Pseudomonas aeruginosa* traced to a scrub nurse with onychomycosis. *Clin Infect Dis.* 2001;33:317–323.

40. Vyhlidal SA. Central services. In: Pfeiffer JA, Ed. *APIC text of infection control and epidemiology, vol. 1.* Washington, DC: Association for Professionals in Infection Control and Epidemiology; 2000:54A-3–54A-4.

41. Rutala WA. APIC guideline for selection and use of disinfectants. *Am J Infect Control.* 1996;24:313–342.

42. Pittet D, Boyce JM. Hand hygiene and patient care: pursuing the Semmelweis legacy. *Lancet Infect Dis.* 2001;9–20.

43. Larson E. Hygiene of the skin: when is clean too clean? *Emerging Infect Dis.* 2001;7:225–230.

44. Pugliese G, Favero MS, Eds. Medical news section: alcohol rubs: CDC's new hand-hygiene guidelines. *Infect Control Hosp Epidemiol.* 2001;22:56–57.

45. Hannigan P, Shields JW. Handwashing and use of examination gloves. *Lancet.* 1998;351:571.

46. Maley MP. Compliance with handwashing. *Infect Control Hosp Epidemiol.* 2000;21:4.

47. Neely AN, Maley MP. Dealing with contaminated computer keyboards and microbial survival. *Am J Infect Control.* 2001;29:131–132.

16 Lessons from "Unexpected Increased Mortality After Implementation of a Commercially Sold Computerized Physician Order-Entry System"

Based on: Sittig DF, Ash JS, Zhang J, Osheroff JA, Shabot MM. Lessons from "Unexpected increased mortality after implementation of a commercially sold computerized physician order-entry system." *Pediatrics.* 2006;118(2):797–801.

16.1 Introduction

This chapter is written in response to a paper by Han et al titled "Unexpected Increased Mortality After Implementation of a Commercially Sold Computerized Physician Order Entry System."[1] The authors are to be congratulated for their courage in bringing their compelling account of computer-based provider order-entry (CPOE) implementation problems to the medical literature, as they tried to interpret their results concerning mortality. Their paper is as much a search for answers as it is a recitation of the shortfalls in their implementation process and computer systems. It is critically important to understand that the types of problems described by Han et al are not limited to their institution. In fact, setbacks and failures in the implementation of clinical information systems (CISs) and CPOE systems are all too common.[2–4] While it is tempting to focus solely on the role of new technology in the problems highlighted by this example, there are also important lessons to be learned about related organizational and workflow factors that affect the potential for danger associated with CPOE implementation.

Many publications have previously addressed the issue of troubled or failed implementations. Han et al's account is unique in that an adverse change in mortality rate was associated in time with CIS and CPOE implementation. We may question the study's methodology and conclude that causality was not proven, yet the assignment of CPOE to a severity-adjusted odds ratio of 3.71 for patient deaths simply cannot be ignored. Regardless of what was or was not proven, if only one unnecessary death were caused by the implementation process or CIS and CPOE modules, that is one too many.

The question that must be asked is this: How can intelligent and well-intentioned leaders at all levels of an institution make the kind of implementation decisions that ultimately place excellent patient care in jeopardy? Clearly, that was not their intent in adopting CPOE, so how could it happen? What is it about CIS and CPOE that makes implementation so risky? Why are these implementations prone to causing emotional distress,[5] rework,[6] delay,[7] user protest,[7] temporary system withdrawal, and later repeat implementation,[8] often at a cost of millions of dollars to the hospital or health system involved? How can institutions avoid these risks and additional costs? These are the questions that demand answers.

We posit that the primary reason why CISs and CPOE are prone to failure is that they have the ability to profoundly alter patient care workflow processes. Although the intent of computerization is to improve patient care by making it safer and more efficient, the side effects and unintended consequences of workflow disruption may, in fact, make the situation far worse.[9] It is important to remember that the manual processes of patient care and documentation in place within an institution have been fine-tuned over long periods of time, usually spanning years or even decades. While paper charting forms, medication ordering, delivery and administration, and processes for patient admission and transfer are appropriate targets for computerization, the transition from manual to computerized methods is notoriously complex. *This is a severely underappreciated fact of CIS and CPOE implementation.* In an ordinary business, employees, clients, revenues, and profits may be adversely affected by computerization. In a hospital, patients and caregivers are at risk. In other words, the stakes are much higher in the healthcare setting.

George Santayana once wrote, "Those who cannot remember the past are condemned to repeat it."[10] Perhaps what we should do in retrospect is to learn from mistakes that occurred in this implementation and others, with the ultimate goal being to help ensure that organizations implementing CIS and CPOE in the future do not fall prey to Santayana's admonition.

16.2 What We Can Learn from This Study

16.2.1 "Hospital-wide implementation of CHP's CPOE system (along with its clinical applications platform) occurred over a 6-day period."

While few organizations have the luxury of pioneering institutions that spent 10 years or more rolling out a CIS and CPOE, attempting such a project in a few days goes beyond challenging and borders on the temerarious.[11] Previous studies have shown that the workflow of clinicians changes significantly following implementation of these types of systems.[12] Given that such a huge change occurs, clinicians must have enough time to adapt to their new routines and responsibilities in a setting that is carefully managed to assure that patient care is not harmed in any way. Contrary to isolated claims of success by several HIT vendors, rapid implementation of any CIS—let alone one that changes the way orders are written and carried out—should not be attempted unless planning has been thorough and resources are abundant.[13,14] Furthermore, time is needed during the implementation process to evaluate whether the changes in workflow are positive or negative, safe or unsafe, and more or less efficient. This cannot be done in a few days, even on a single ward. Experience with countless prior CIS and CPOE implementations has shown that not all changes in workflow represent improvements.[15,16]

16.2.2 "After CPOE implementation, order entry was not allowed until after the patient had physically arrived to the hospital and been fully registered into the system."

While accurate patient registration is clearly important to patient safety, the care and treatment of a severely ill patient should never be made to wait for a computer system. Analysis by multidisciplinary teams regarding the workflows that were successful prior to system implementation should have led clinicians and system administrators to develop a means to allow clinicians to continue to treat patients in the best way possible. This might require using old-fashioned paper orders in emergency situations, with subsequent entry into the CPOE system after the patient is stable. Under no circumstances can the care of a patient be subordinated to the idiosyncrasies of a computer system.

16.2.3 "As part of CPOE implementation, all medications, including vasoactive agents and antibiotics, became centrally located within the pharmacy department."

The relocation of all medications, including vasoactive drugs used in the ICU, to a central pharmacy—even if it were done for administrative reasons without implementation of a CIS and CPOE—could account for many of the adverse effects noted in Han et al's study. Considering that the hospital was already undertaking a huge disruptive organizational change affecting every caregiver in the institution, it was unfortunate that it would also try to institute a significant policy change regarding pharmacy workflow to accommodate CPOE more effectively. The additive effects of the CIS implementation, CPOE implementation, and pharmacy centralization could have been predicted to dramatically slow the delivery of drugs to all patients. Han et al's selection of interfacility transport patients as the patient population for analysis probably magnified the ill effects, as these patients can be predicted to be more severely ill on admission than other patients. Piggybacking organizational changes with significant potential for adverse workflow effects onto a CIS/CPOE implementation should be avoided if at all possible. CIS and CPOE are disruptive in and of themselves; inter-related workflows should be enhanced prior to implementation where possible, or at least remain stable through the implementation period, to minimize this disruption.

Many hospitals use small "tests of change" on a single hospital unit to evaluate a new care process for both efficacy and potential adverse effects or unintended consequences before rolling out that process on a larger scale. If pharmacy centralization had been evaluated in a single ICU in advance of CIS and CPOE implementation, it is likely that the operational problems described by Han et al would have been appreciated so that appropriate solutions could have been put into place.

16.2.4 "Because the pharmacy could not process medication orders until they had been activated, ICU nurses also spent significant amounts of time at a separate computer terminal and away from the bedside."

The diversion of ICU nurses from bedside care was clearly an unintended consequence of computerization at Han et al's facility. Careful sociotechnical analysis is required before clinical systems are implemented to ensure that caregivers can do their basic job at least as well and as safely as they could before computerization. Whenever an organization commits to moving from a well-honed manual care delivery system to a new computer-based model, the organization should carefully review and modify all applicable practices, procedures, policies, and bylaws. Mock use, full dress drills, and trial use on individual patients and wards should precede wider implementation of the change. The role of the computer in health care is not to ensure that rules and regulations that had never been completely followed are now rigorously followed. Rather, computerization highlights the need for review, careful consideration of purpose, and clear definition of intended policies. Allowing a computer to enforce rules and regulations without first working through all implications and potential unintended consequences for patient care is a prescription for disaster.

16.2.5 "After CPOE implementation, because order entry and activation occurred through a computer interface, often separated by several bed spaces or separate ICU pods, the opportunities for such face-to-face physician–nurse communication were diminished."

Clear, two-way, face-to-face communication is the hallmark of high-quality, collaborative patient care. Assuming that ambiguities in the treatment process are removed because all the orders are now legible and available in a central database is inappropriate and potentially dangerous. As the importance and complexity of the information to be communicated increases, the necessity of face-to-face communication increases dramatically. Careful sociotechnical evaluation of the proposed computerization-related changes, or even trial use on a single ward, probably would have brought to light the need for better system design. In addition, recent advances in the capability and utility of mobile terminals, tablet computers, and other devices such as hands-free, wireless communication systems[17] may allow caregivers to remain in personal contact while doing their computer work. In fact, careful attention to these details has been shown to bring care teams together and make them more—not less—effective.[18,19] Computer systems need to be designed and implemented in such a way as to foster appropriate levels of communication, not hinder it.

16.2.6 "This initial time burden seemed to change the organization of bedside care. Before CPOE implementation, physicians and nurses converged at the patient's bedside to stabilize the patient. After CPOE implementation, while one physician continued to direct medical management, a second physician was often needed solely to enter orders into the computer during the first 15 minutes to 1 hour if a patient arrived in extremis."

As this quote suggests, the consequences of CPOE were not appreciated at Han et al's facility until after its implementation. Doubling the physician workload, while slowing the delivery of life-saving medications, treatments, and diagnostic studies, could not have been the original intent of the CPOE implementation. Careful pilot studies could have revealed these issues so that solutions could have been devised before hospital-wide implementation. While several studies have shown a small but significant increase in the time required on the part of clinicians to enter orders using a computer system, no one has ever documented a "doubling" of physician workload (see Poissant et al's review of several studies in this area[12]).

16.2.7 "The physical process of entering stabilization orders often required an average of ten 'clicks' on the computer mouse per order, which translated to ~1 to 2 minutes per single order as compared with a few seconds previously needed to place the same order by written form. . . . However, no ICU-specific order sets had been programmed at the time of CPOE implementation. . . ."

Methods of entering frequently occurring orders should be as easy and fast as on paper, especially for sets of orders, while providing the added benefits of 100% legibility, instantaneous transmission to the ancillary department, dose-range checking, and potential drug, laboratory, and condition interaction checking.[20] Organizations must take the time to implement validated standard order sets for routinely occurring critical conditions to speed the ordering and care process.[21] Simply training users to overcome a steep learning curve and the time-consuming process of entering many individual orders in the midst of critical patient care is not an optimal—or even effective—solution. Therefore, organizations must work to ensure that the clinical content (e.g., order sets), default settings, and anticipated screen flows are designed, implemented, and tested so as to optimize speed, usability, and patient safety. This effort may require trial use of the system on one ward, or a few well-defined and well-staffed wards, by a variety of users and over a prolonged period of time; it cannot be done in a few days. A more reasonable estimate of the amount of time to fully develop and vet clinical policies and order sets, and configure, test, and implement CPOE systems is 1–3 years.[22]

16.2.8 "Because the vast majority of computer terminals were linked to the hospital computer system via wireless signal, communication bandwidth was often exceeded during peak operational periods, which created additional delays between each click on the computer mouse."

Technical issues such as the one described by Han et al can also be anticipated and tested in advance. On top of all the other process and workflow changes involved in CPOE implementation, inadequate or unreliable computing capacity can be particularly frustrating to clinicians and other end users. Testing a new CIS under peak load conditions is an important task that cannot be overlooked.

16.3 Conclusion

Although it is not clear whether the increase in mortality rate reported by Han et al was due directly to the CPOE implementation or other concomitant organizational and system changes, the CPOE implementation may well have been responsible, and we applaud the authors for reporting their findings and their problems with implementation. Although it is easy to criticize organizations for reporting implementation decisions that in retrospect appear flawed, we must respect, appreciate, and encourage other institutions to share their experiences so that everyone can learn from them. Likewise, whether or not the technology has a direct role in adverse effects from a specific deployment, we can take the opportunity from case studies like this one to learn how to develop better systems.

The complexity of the decisions that must be made by CIS implementation and management teams demands iterative ongoing dialogue and feedback over time. There is no substitute for careful workflow and sociotechnical analysis, and beyond those considerations there is no substitute for trial or pilot use of a system to uncover hidden flaws, unintended consequences, and adverse effects.

One must avoid the inclination or temptation to blame the adverse effects noted in Han et al's paper solely on the particular CIS or CPOE system used. This would be the equivalent to stating that a particular brand of tool from a hardware store was unsafe because an injury occurred while misusing it.

To return to the central question, how can well-intentioned organizations avoid these problems? Beyond the solutions noted earlier, several collective publications on CIS and CPOE implementation are available to guide institutions in designing a safe and effective process.[23–26] The advice in these guides, along with a careful evaluation of caregiver workflows and trial implementation in limited hospital areas, should allow safe implementation for everyone involved, especially patients.

A very important lesson in Han et al's paper is the need to measure overall hospital mortality and adverse event rates when implementing major new systems. Indeed, these mortality rates could go up even when rates of adverse drug events (ADEs) go down, as concomitantly reported from Han et al's institution.[27] However, these particular findings need to be interpreted cautiously because they relied on "self-reported" ADEs that may have little to do with the true underlying rate of adverse effects.[28] The efficacy of CPOE systems has traditionally been measured in terms of ADE rates; Han et al have reminded us that we must consider the larger scope of patient outcomes if we are to accurately evaluate safety and efficacy.

We believe that the problems observed in Han et al's and other CPOE deployments can be overcome by systematically developing and applying human-centered design, implementation, and evaluation methods adapted to point-of-care clinical information systems. Such a systematic approach, as advocated by experienced practitioners in the field of medical informatics, has been achieved in aviation, the military, nuclear power, and the consumer software industry. It can be—indeed, it must be—achieved in health care as well.

16.4 References

1. Han YY, Carcillo JA, Venkataraman ST, et al. Unexpected increased mortality after implementation of a commercially sold computerized physician order-entry system. *Pediatrics*. 2005;116(6):1506–1512.

2. Southon G, Sauer C, Dampney K. Lessons from a failed information systems initiative: issues for complex organisations. *Int J Med Inform.* 1999;55(1):33–46.

3. Goddard BL. Termination of a contract to implement an enterprise electronic medical record system. *J Am Med Inform Assoc.* 2000;7(6):564–568.

4. Wager KA, Lee FW, White AW. *Life After a Disastrous Electronic Medical Record Implementation: One Clinic's Experience.* Hershey, PA; Idea Group Publishing: 2002.

5. Sittig DF, Krall M, Kaalaas-Sittig J, Ash JS. Emotional aspects of computer-based provider order entry: a qualitative study. *J Am Med Inform Assoc.* 2005;12(5): 561–567.

6. Payne TH, Hoey PJ, Nichol P, Lovis C. Preparation and use of preconstructed orders, order sets, and order menus in a computerized provider order-entry system. *J Am Med Inform Assoc.* 2003;10(4):322–329.

7. Massaro TA. Introducing physician order entry at a major academic medical center: I. Impact on organizational culture and behavior. *Acad Med.* 1993;68(1):20–25.

8. Scott JT, Rundall TG, Vogt TM, Hsu J. Kaiser Permanente's experience of implementing an electronic medical record: a qualitative study. *BMJ.* 2005;331(7528):1313–1316.

9. Ash JS, Berg M, Coiera E. Some unintended consequences of information technology in health care: the nature of patient care information system-related errors. *J Am Med Inform Assoc.* 2004;11(2):104–112.

10. Santayana G. *Life of Reason, Reason in Common Sense.* New York: Scribner's; 1905:284.

11. McDonald CJ, Overhage JM, Tierney WM, et al. The Regenstrief medical record system: a quarter century experience. *Int J Med Inform.* 1999;54(3):225–253.

12. Poissant L, Pereira J, Tamblyn R, Kawasumi Y. The impact of electronic health records on time efficiency of physicians and nurses: a systematic review. *J Am Med Inform Assoc.* 2005;12(5):505–516.

13. Cerner press release. CPOE improves patient safety at top-ranked pediatric hospital. Available at: http://www.cerner.com/public/NewsReleases_1a.asp?id=257 &cid=4668. Accessed July 14, 2009.

14. Baker ML. Management plays key role in success of electronic patient record system. *Ziff Davis Internet.* Available at: http://www.eweek.com/article2/ 0,1895,1600999,00.asp. Accessed July 14, 2009.

15. Weingart SN, Toth M, Sands DZ, Aronson MD, Davis RB, Phillips RS. Physicians' decisions to override computerized drug alerts in primary care. *Arch Intern Med.* 2003;163(21):2625–2631.

16. Nebeker JR, Hoffman JM, Weir CR, Bennett CL, Hurdle JF. High rates of adverse drug events in a highly computerized hospital. *Arch Intern Med.* 2005;165(10): 1111–1116.

17. Breslin S, Greskovich W, Turisco F. Wireless technology improves nursing workflow and communications. *Comput Inform Nurs.* 2004;22(5):275–281.

18. Ash JS, Stavri PZ, Dykstra R, Fournier L. Implementing computerized physician order entry: the importance of special people. *Int J Med Inform.* 2003;69(2–3): 235–250.

19. Reddy M, Pratt W, Dourish P, Shabot MM. Sociotechnical requirements analysis for clinical systems. *Meth Info Med.* 2003;42:437–444.

20. Bates DW, Boyle DL, Teich JM. Impact of computerized physician order entry on physician time. *Proc Annu Symp Comput Appl Med Care.* 1994;996.

21. Ali NA, Mekhjian HS, Kuehn PL, et al. Specificity of computerized physician order entry has a significant effect on the efficiency of workflow for critically ill patients. *Crit Care Med.* 2005;33(1):110–114.

22. Payne TH, Hoey PJ, Nichol P, Lovis C. Preparation and use of preconstructed orders, order sets, and order menus in a computerized provider order-entry system. *J Am Med Inform Assoc.* 2003;10(4):322–329.

23. Lee F, Teich JM, Spurr CD, Bates DW. Implementation of physician order entry: user satisfaction and self-reported usage patterns. *J Am Med Inform Assoc.* 1996;3(1):42–55.

24. Osheroff JA, Pifer EA, Teich JM, Sittig DF, Jenders RA. *Improving Outcomes with Clinical Decision Support: An Implementer's Guide.* Chicago: Healthcare Information and Management Systems Society; 2005.

25. Metzger J, Fortin J. *Computerized Physician Order Entry in Community Hospitals: Lessons from the Field.* Oakland, CA: California Healthcare Foundation; 2003.

26. Drazen E, Kilbridge P, Turisco F. *A Primer on Physician Order Entry.* Oakland, CA: California Healthcare Foundation; 2000.

27. Upperman JS, Staley P, Friend K, et al. The impact of hospital-wide computerized physician order entry on medical errors in a pediatric hospital. *J Pediatr Surg.* 2005;40:57–59.

28. Classen DC, Pestotnik SL, Evans RS, Burke JP. Computerized surveillance of adverse drug events in hospital patients. *JAMA.* 1991;266(20):2451–2847.

Index

A

Abbreviations, 19, 45
Accuracy of data, 125, 127–128, 130–131
Administrative needs, 148
Administrative staff, 30
Advanced life support situations, 119
Albany Medical Center, 12
Alerts
  alert fatigue, 57, 72, 79, 81, 109, 111
  contextual relevancy, 95
  contradictory advice offered by, 107
  decision support implementation, 153
  poorly designed, 47
Allen, S.I., 19
Amalberti, R., 82, 83
Ambiguity, 8
Ambulatory care units, 119
American Hospital Information Management
  Association, 40
Anderson, J.G., 20, 21
Anticipated consequences survey, 181–189
  methods, 183
  perceived advantages/disadvantages of
    CPOE, 184
  rapid ethnographic assessment methods,
    182
  recommendations, 187–188
  report to CPOE leadership, 185–187
  results, 183–186
  study limitations, 186
  survey sites, 182–183
  update, 188
Antimicrobial soaps, 207
Artificial intelligence-based computer system,
  22
Ash, J.S., 67, 177
Auto-complete features, 109
Autonomy, loss of, 43, 80–81
Aydin, C.E., 177

B

Backup systems, 39, 44, 132, 137–138
Bellotti, V., 59
Berg, M., 54, 62
Blood transfusions, 22–23
Bone marrow transplant units, 119
"Bottom-up" feedback, 122
Bridgers, 167, 173–176
  key attributes of, 176
  skills and training for, 175–176
Brown, C.S., 19
Bures, S., 202
Bylaws, 78

C

CARE rules, 7
Cefuroxime order-entry screen, 9f
CEO (chief executive officer), 167–168, 178
Certification Commission for Healthcare
  Technology, 56
Champions, 168, 170–171
Chief executive officer (CEO), 167–168, 178
Chief information officer (CIO), 168–169, 178
Chief medical information officer (CMIO),
  169–170, 178
Childs, B.W., 18
CIO (chief information officer), 168–169, 178
Cleaning and disinfecting, 206
Clerical tasks, 157
Clinical activity sequencing, 57
Clinical application coordinators, 173
Clinical decision-support systems (CDS), 8,
  10–11, 105–114
  alert fatigue, 57, 72, 79, 81, 109, 111
  algorithm-based rules, 107
  defined, 105
  elimination/shifting of human roles with, 106
  error sources, 109
  homegrown vs. commercial, 110

inadequacy of, 108
knowledge management structure and, 110–111
overdependence on, 128–129, 185
questions, 111–113
rigidity of, 108
situation awareness and, 60
unintended adverse consequences of, 105–113, 128–129
updating, 106–107, 109, 110–111
wrong or misleading content in, 107–108
Clinical end users, 30
Clinical information system. *See also* Rapid assessment of CIS interventions
committee power and, 81–82
education in, 96–97
improving with positive feedback, 73–74
inflexibility, 92–93
medical errors and (*See* Medical errors)
"read-only" version of, 133
Clinical leadership considerations, 155–156
Clinical needs, 148
Clinical software systems, 38–39, 46
Clinical systems committees, 172–173
Clinical workflow, 53–65
alteration of, 63, 216
analysis, 220
changing practice patterns, 12–13
CPOE implementation and, 148, 151
defined, 53–54
interruptions, and error potential, 89–90, 97
labor shortages, 148
power structure shifts, 78
questions, 63–65
sociotechnical conceptualization of, 54
sociotechnical system evaluation, 54–55
unintended consequences and, 37–38, 55–62
Clinician autonomy, 43, 80–81
Clyman, J., 17
CMIO (chief medical information officer), 169–170, 178
Coalition power, 81–82
Code Blue situations, 119, 138
Coding updates, 106–107
Collen, M.F., 3, 199
Colonization, 201
Co-management, 118
Communication, 115–123
anticipated consequences of CPOE and, 185, 188

bidirectional administration-staff communication, 120–121
in CPOE increased mortality study, 218
diminishing of personal nature of, 116
disturbance of doctor-nurse, 116–117
downtime communication procedure, 133
errors, 92–96
leadership skills in, 177
medical teams and, 118
medication orders, 117–118
misrepresentation of as information transfer, 94–96
multiple provider, 61
nonverbal, 118
patient communication, 118
questions, 123
reciprocal impacts of CPOE on, 115–116
situation awareness and, 61
time considerations, 151
time/location variations and, 118–121
"top-down," 122
untoward changes in patterns/practices of, 41, 46
Community Memorial Hospital (Toms River, NJ), 16
Computer-based direct physician order entry (CPOE). *See also* Anticipated consequences survey; Emotional response to CPOE; Implementation (CPOE); Unintended consequences
benefits of, 8
changing practice patterns, 12–13
clinical decision support, 10–11, 47, 128–129
cost-conscious decision making through, 9–10
data accuracy, 125, 127–128, 130–131
ergonomic and human/computer interaction issues, 55–56
errors related to, 42–43
facilitation of, 21–23
goals of, 11–12
implications for future systems, 23–24
inadequate institutional policies, 14
increased mortality study, 215–220
ineffectiveness highlighted by, 61–62
informational text inserted into, 10
outpatient prescription-writing systems, 18–19
process improvement via, 8–9
radiologic procedures and, 21–22

rationale for direct, 7–11
reminders generated at time of, 11
shifting care team roles, 13
situation awareness reduction and, 59–61
sociologic barriers, 11–14
system design, 17–21, 58–59
system downtime, 47, 125, 126–127,
   129–130
traditional teaching changes, 13–14
transfusion orders and, 22–23
work pace/sequence/dynamics changes and,
   56–58
Computer hardware
medical errors and, 86
platforms, 38–39, 46
transmission of pathogen transmission via
   (*See* Microbiologic and infection control)
Computer-human interaction issues, 55–56
Computer software
application demands, 38–39, 46
medical errors and, 86
Computer-to-staff ratios, 56
Computer workspace limitations, 55
Confidentiality, 140
Consultant orders, 18
Contingencies, 88, 94
Control, loss of, 79–80
Cooper, R.B., 177
Cost considerations, 150–151
infrastructure expense, 168
laboratory tests and, 10
savings, 9
system downtime, 129
Countersignature orders, 18
Curmudgeons, 172
Customization, 157, 158

D

Data accuracy, 125, 127–128, 130–131, 157
medication reconciliation issues, 108
mistrust about data provenance, 108
"write" data, 128
Data backup, 156
Data entry structure, 18, 90
Data placement, 42
Decision making
cost consciousness of, 9–10
optimization of physician time and, 11
Decision support, 8, 47
flexibility in task completion, 158

implementation, 153
overload, 95–96
physician autonomy and, 81
power structure shifts with implementation
   of, 78–79
Desktop operating systems, 131
Devine, J., 202
Diffusion of innovations theory, 30
Diffusion (Rogers), 30
Digital information resources, 14
DiPalma, C., 83
Disaster recovery, 156
Disinfecting, 206
Doctors. *See* Physicians
Documentation. *See* Persistent paper
Dourish, P., 59
Downtime
communication, 133
drills, 130, 132
emergency kit, 132
notification procedures, 133
order procedures, 18
paper backups during, 139
preparation class, 132
procedure development, 47, 188
Downtime committee, 133
Drugs. *See* Medications
DxCON, 22

E

Eclipsys Sunrise system, 182
Education and training, 154
Efficiency needs, 148
Efficiency perception, 153
El Camino Hospital (Mountain View, CA),
   4, 182, 183, 188
Electronic signatures, 17, 141
Embedding issues, 152
Emergency power backup system, 132
Emergency room environment, 119
Emotional response to CPOE, 67–75
cognitive and physical abilities and,
   67–68
emotions defined, 67
negative emotions, 41–42, 67, 71–72, 74
neutral emotions, 70–71, 74
positive emotions, 68–70
questions, 74–75
taxonomy of terms, 68, 68t
unintended consequences, 68–72

Epic Systems, 182
Ergonomics, 55–56, 137
Errors. *See* Medical errors
ESPRE, 22
Ethnography, 192–193
Expert power, 77, 78
Expert systems, 21–23

F
Feedback, 155
Field manual development, 193–194
Flowsheets, 136
Forsythe, Diana E., 165–166, 177
Free-text fields, 45
French, J.R.P., 77

G
Gardner, R.M., 6
Garside, D.B., 19
Gloving, 206–207
Guideline for Hand Hygiene, 207

H
Han, Y.Y., 215
Handwashing, 206–207
Handwritten orders, 8
Hardware. *See* Computer hardware
Harper, R., 139
Healthcare reform, 23–24
Health Insurance Portability and Accountability
    Act (HIPAA), 140
Hectic environments, 119–121
HELP clinical information management system,
    6–7, 22
HIPAA, 140
Hodge, M.H., 9
Horizon Expert Orders (McKesson), 183
Hospital discharges, 138
"Hot site," 132
Human-computer interaction issues, 55–56
Hygiene, 206–207

I
Imaging orders, 117
Implementation (CPOE), 3–7, 147–163.
    See also Special people
    benefits, 8–11
    costs of, 150–151
    foundations needed prior to, 148–150

highest-level considerations, 158–159,
    161–162
integration/workflow/healthcare
    considerations, 151–152
leadership, 148, 149, 163
learning/evaluation/improvement
    considerations, 161–162
localization, 160
logistic challenges in, 14–17
lower-level considerations, 159–160
mid-level considerations, 159
motivation for, 147–148
order input time, 17
phasing, 14–15
promoting usage of, 152
Regenstrief Medical Center, 7
roles, 160
strategies, 162
support of cost-conscious decisions with,
    9–10
technical considerations, 156–158
Technicon Data Systems, 4–5
terminal numbers and location, 15–16
training/support/help considerations, 16,
    160–161
value to users/decision-support systems,
    152–154
vision/leadership/people considerations,
    154–156
workflow issues, 148, 216–217
Infection, 200, 201–202, 209. *See also*
    Microbiologic and infection control
Influence, 77
Informatics education, 96–97, 130
Information accuracy. *See* Data accuracy
Information power, 77, 78
Information retrieval, 152
Information security officer, 83
Information technology, 30, 79, 80, 82, 177
Infrastructure, 149, 157, 168
Innovation, 30
Input requirements, 53
Institute of Medicine, 85, 199
Intensive care units, 119
Interaction of drugs, 11
Interface design usability, 55–56, 157
Interruptions, and potential for error, 89–90, 97
Isaacs, D., 202
Ivey, J.C., 204

J

JCAHO guidelines. *See under* Joint Commission
Joint Commission
  guidelines, 6
  updating of CDS content and, 107
Juxtaposition errors, 42

K

Kaiser Permanente Sunnyside Hospital
    (Portland, OR), 182–183, 188
Kaplan, B., 54, 191
Kaufman, D.R., 177
Kawahara, N.E., 9
Keyboards and infection risk, 203–204,
    208–209. *See also* Microbiologic and
    infection control
Knowledge management, 110–111
Kochi Medical School Hospital (Japan), 8

L

Laboratory orders, 117
Labor shortages, 148
LDS Hospital (Salt Lake City, UT), 6–7
Leadership, 154–155. *See also* Special people
  administrative, 167–170
  clinical leadership level, 167, 170–173
  CPOE implementation, 148, 149, 163
Leapfrog Group, 148
Learning organization, 149
Legal documentation, 46
Legitimate power, 77, 78
Lepage, E.F., 6, 22
Levine, H.S., 20
Levit, F., 19
Light-pen technology, 18
Lorenzi, N.M., 177
Luff, P., 139

M

Massaro, T.A., 5, 13, 17
Masterton, R.G., 202
McDonald, C.J., 7, 15, 71
Mechanic, D., 83
Medicaid, 147–148
Medical errors, 42–43
  clinical information system-related, 85–99
  cognitive overload, 90–92
  in communication/coordination process,
    92–96

hardware problems and software bugs,
    86–87
human-computer interfaces, 89–90
identifying/exposing, 98–99
in information entering/retrieving, 88–92
main categories of, 88
overreliance on clinical decision support and,
    128
"silent" errors, 88, 98
study background and methods, 87–88
system design, 97–98
system downtime and, 127
in U.S. medical care system, 85
"Medical gopher," 7
Medical information management system
    objectives, 3
Medications
  administration prompts, 10
  alerts, 79, 112–113
  altering timing of dosage, 58
  calculations for dosage, etc, 10
  communication and orders for, 117–118
  cost-effective options, 9
  cost information, 79
  in CPOE increased mortality study, 217–218
  delay in administration of, 57–58
  drug-drug interactions, 11, 59, 109
  drug-recall information, 10
  errors in, 5, 85, 86, 93, 96
  ordering, CDS and, 106, 109
  outpatient prescription-writing systems,
    18–19
  reconciliation issues with CDS, 108
  urgency of orders and, 93
Member checking, 196
Methodist Hospital of Indiana, 21
Microbiologic and infection control, 199–211
  computers as microbial reservoirs, 201–202,
    203t, 209
  microbial survival and transfer, 204–205
  nosocomial infections, 201–202
  patient risk level at each computer location,
    208
  patients at risk of infections, 200, 200t
  personnel, 208
  principles of, 205–209
  questions, 210
  steps preceding an infection, 199–201, 200f
"Midnight problem," 94

Missouri Automated Radiology System (MARS), 21
Modifications, 160
Morelos-Borja, H., 67

N
National Health Policy forum, 58
Neely, A.N., 202
"New Bottles, Old Wine: Hidden Cultural Assumptions in a Computerized Explanation System for Migraine Sufferers" (Forsythe), 165–166
New York University Medical Center, 5
Nonverbal communication, 118
Noskin, G.A., 205
Nosocomial infections, 201–202, 209
Nurses
    communication with doctors, 116–117
    CPOE design and, 58–59
    power shifts and, 80, 82–83

O
Observation guide, 193, 196
Obstetrics, 61
Off-site backup, 132
Ogura, H., 16
Opinion leaders, 171–172
Order-entry pathways, 12–13
Ordering time, 151
Order process
    levels of healthcare workers and, 59
    loss of feedback, 95
    order modifications, 18
    quality assurance monitors and, 8
    timing of, 60
Order sets, 19–21, 20*t*, 39, 154
Order standardization, 44
Order terminology, 80
Outpatient settings, 88
Output requirements, 53
Overcompleteness of reports, 91
Overview loss, 90

P
Paper records, 152. *See also* Persistent paper
Passwords, 83
Patel, V.L., 177
Patient care
    benefits, 154

communication, 118
flowsheet, 136
patient flow and transfer, 93–94
patient privacy, 83
Persistent paper, 40, 46, 135–142
    dual documentation, 138
    hybrid documentation, 140
    negative aspects of, 140
    positive aspects of, 139
    questions, 142
    rationale for, 136–139
    regulatory factors, 138–139
Personal order sets, 20
PHOENIX expert system, 21–22
Physician order entry. *See* Computer-based direct physician order entry (CPOE)
Physicians
    autonomy of, 43, 80–81
    communication with nurses, 116–117
    optimization of physician time, 11
    project leaders for CPOE implementation, 156
Pick lists, 42
POET research team, 183
Policy changes, 78
Power shifts, 43–44, 47, 77–79, 186
    forced work distribution, 78
    mandated changes for safety pursuits, 78–79
    power defined, 77
    questions, 83–84
Preadmission orders, 18
Prescriptions, 18–19. *See also* Medications
Problem-solving, 172
Process improvement, 8–9
Providence Portland Medical Center (Portland, OR), 183, 188

Q
Quality-assurance monitors, 8
Quick ethnography, 192

R
Radiologic procedures, 21–22
Rangel-Frausto, M.S., 204–205
Rapid Assessment, Response, and Evaluation Project (RARE), 192
Rapid assessment of CIS interventions, 191–196
    data analysis, 195–196

data collection, 195
data management, 195
field manual development, 193–194, 193t, 194t
lessons learned, 196
methodological approach selection, 192–193
results, 196
site visit preparation, 194
subject selection, 194–195
Rapid Assessment Process (RAP), 192–193, 196
Raven, B., 77
Reactivation procedures, 132, 188
Reddy, M.C., 82
Referent power, 77, 78
Regenstrief Medical Center, 7
Regulatory environment, 138–139, 147–148
Related orders, automatic generation of, 8
Remote access, 157
Replacement of CPOE systems, 157
Reports, overcompleteness of, 91
Research field manual, 193
Resources, 149
Response time, 151
"Revenge effects" of technology, 39
Reward power, 77
Reynolds, M.S., 9
Riley, R.T., 177
Risk analysis, 157, 215–220
Rogers, Everett, 30
Role boundaries, 82

S

Safety. *See also* Microbiologic and infection control
mandated changes for, 78–79
oversight committee, 83
Saleem, J.J., 82
Santayana, George, 216
Schroeder, C.G., 16
Security of information, 83, 140, 156
Sellen, A., 139
Site inventory
assessment tool, 193, 193t, 196
profile, 194
Sittig, D.F., 41
Situation awareness, 59–61
Soaps, 207
Social construction of technology theory, 126

Social organization of medical work, 88
Sociotechnical systems, 54–55, 220
Software. *See* Computer software
Space limitations, 55
Spackman, K.A., 22
Special people, 165–179
administrative leadership, 167–170
clinical leadership level, 167, 170–173
need for, 177
overlapping roles of, 166–167
questions, 179
Spectra, 3
Standardization of orders, 44
Sterilization (infection control), 206
Sticky notes, 137, 139
Storm, C., 68
Storm, T.A., 68
Stressors, 72
Subtask sequencing, 53
Support staff level (bridgers), 167, 173–176
Suspended orders, 18
System design
capabilities/policies required, 17–18
data entry methods, 18
errors, 97–98
order sets, 19–21
outpatient prescription-writing systems, 18–19
System downtime, 125, 126–127, 129–130, 188
costs of, 129
paper backup systems for, 130, 139

T

Task completion, 158
Teamwork, 156, 173
Technical considerations (CPOE implementation)
flexibility in task completion, 158
strategic-level, 156–157
user considerations, 157
Technicon Data Systems, 4–5
Technological determinism theory, 125–126
Technology
clinical decision support, 128–129
data accuracy, 127–128, 130–131
overdependence on, 44–45, 47, 125–133
"revenge effects" of, 39
sociotechnical system evaluation, 54–55
system downtime, 125, 126–127, 129–130

Technology solution, 54
Tests
    cost information on, 10, 79
    reducing number of, 10
Test system for patient names, 43, 99
Tierney, W.M., 7, 8, 10, 17
Time-motion study, 17, 54
To Err Is Human (Institute of Medicine), 148
Training, 16, 154, 160–161
    bridgers and, 174–175
    models and methods, 16
    recommendations, 188
Transfer of patients, 93–94
Transfusion Advisor (TA), 22
Transfusion orders, 22–23
Translators, 173
Trust, 149

U
"Unexpected Increased Mortality After
    Implementation of a Commercially Sold
    Computerized Physician Order-Entry
    System" (Han et al), 215–220
    bedside care, 218–219
    communication issues, 218
    computing capacity, 219
    medications and, 217–218
    order entry issues, 217, 219
    time table, 216–217
Unintended consequences. *See also* Anticipated
    consequences survey; Persistent paper
    categories of, 30–32, 31t, 33–36t, 37–45
    classification of, 29
    communication patterns/practices and, 41, 46
    decision support systems, 45, 47
    defined, 30
    demands for system changes, 38–40, 46
    discovering causative factors leading to, 48
    to emotional response to CPOE, 47, 68–72
    frequency of occurrence, 32t
    key points, 29
    more/new work for clinicians, 32, 37, 45–46

negative emotions, 41–42
new kinds of errors, 42–43, 47
overdependence on technology, 44–45, 47,
    125–133
power structure changes, 43–44, 47, 77–79
recommendations to increase awareness of,
    187–188
related to clinical decision-support systems,
    105–113, 128–129
related to clinical workflow, 37–38, 46,
    55–62
theoretical framework, 30
unintended adverse consequences (UACs),
    29–30
University of Minnesota Hospital and Clinic, 22
University of Virginia, 5, 16, 20
User considerations, 157
User-friendly menus, 18

V
Value proposition, 152–153
Vendor readiness, 150, 159
Verbal orders, 13, 117
"Versioning," 140
Vision, 149, 154, 177

W
Ward clerks, 80
Weir, C., 177
"Wet signature" requirements, 141
Workarounds, 93, 159
Workflow. *See* Clinical workflow
Work pace alterations, 56–57
Work structure changes, 151
"Write" data, 128

Y
Yale-New Haven Hospital, 17
Yale University, 22

Z
Zmud, R.W., 177